Australia in Space

*To every Australian who wants to reach for the stars
and every Australian working to put Australia in space*

KERRIE DOUGHERTY

Australia in Space

A History of a Nation's Involvement

LEFT *The Hubble Space Telescope being serviced by two astronauts as the Space Shuttle crosses South Australia in December 1993. (Photo courtesy of NASA)*

Copyright © 2017 Space Industry Association of Australia Limited.
Copyright © of photographs held by acknowledged sources where indicated.

All rights reserved. Except for any fair dealing permitted under the Copyright Act, no part of this book may be reproduced by any means without prior permission. Inquiries should be made to the publisher.

National Library of Australia Cataloguing-in-Publication
Creator: Dougherty, Kerrie Anne, author.

Title: Australia in space : a history of a nation's involvement / Kerrie Anne Dougherty.

ISBN: 9781925309645 (paperback)
           9781925309652 (hardback)
           9781925309669 (ebook: ePub)
           9781925309676 (ebook: pdf)

Notes: Includes index.

Subjects: Astronautics--Australia--History.
Outer space--Exploration--Australia--History.

Space Industry Association of Australia Limited
ACN 613 961 005
c/- Nova Systems
London Road 27-31
Mile End SA 5031
Australia
ABN 67 613 961 005

Making a lasting impact
An imprint of the ATF Publishing Group
ATF (Australia) Ltd
PO Box 504 Hindmarsh
SA 5007
ABN 90 116 359 963
www.atfpress.com

Cover Photograph: Courtesy of NASA. Image by Robert Simmon and Reto Stöckli

Graphic Design & Layout: Lydia Paton
Body Font: Bembo Std 12pt
Title & Heading Fonts: Futura Lt BT & Bembo Std

Contents

Contents	i
Preface *Michael Davis*	iii
Foreword *Andrew Thomas*	iv
Chapter 1 Countdown: Australia's First Rocketeers	1
Chapter 2 Ignition! The Origins of Space Activity in Australia	9
Chapter 3 Liftoff: Going Up Space from Down Under	29
Chapter 4 Keeping on Track: Space Tracking From Australia	51
Chapter 5 The High Ground: Using Space for Defence and National Security	73
Chapter 6 An Infrastructure in Orbit: Satellites at Your Service	91
Chapter 7 Reaching for the Stars: Australian Space Science and Engineering	115
Chapter 8 Groundwork: Australian Space Industry	139
Chapter 9 Ground Control: Australian Space Policy	163
List of Acronyms	178
Timeline of Space Events	182
About the Author	187
Acknowledgements	188
Index	189

Preface

The Space Industry Association of Australia is delighted to partner with ATF Press in the publication of *Australia in Space* by Kerrie Dougherty.

Under the title *Space Australia* a first edition of this book was published in 1993 by the Powerhouse Museum, Sydney, and we are grateful to the Museum and Space Australia's original co-author, Mr Matthew L James, for their kind permission to publish a new edition. Space Australia recorded the remarkable achievements and challenges faced by the pioneers of our industry. This edition augments that history with the further achievements of the past 25 years. In this period, the private sector has played an increasingly important role and we find that Australia is now at the forefront of the Space 2.0 phenomenon in which young Australian entrepreneurs are developing ambitious plans to build global systems and compete on the world stage.

Left The Nullarbor Plain seen from orbit. This photo was taken from the Space Shuttle Endeavour while docked to the Mir space station during the STS 89 mission that delivered Australian astronaut Andy Thomas to the Russian outpost for a long-duration spaceflight. (Photo courtesy of NASA)

We have much to celebrate in the rich history of Australia in space. This book is not only a carefully researched record of the highlights of our Australian space history, it is also a beautifully written and illustrated work dedicated to the many participants in the Australian space adventure over a period of 70 years since the founding of Woomera. We now have a complete history, rewritten 50 years after the launch of Australia's first and most notable satellite launch. The WRESAT story continues to inspire us.

We are very fortunate that Kerrie Dougherty agreed once again to take on this project, giving us the benefit of her long involvement and research and her clear insights. As the peak voice for the space industry in Australia we are proud that we have been able to make this contribution and we hope that this story will help to educate and inspire the next generation and encourage further exploits and achievements for Australia in space.

Michael Davis
Chair, Space Industry Association of Australia
Adelaide, Australia

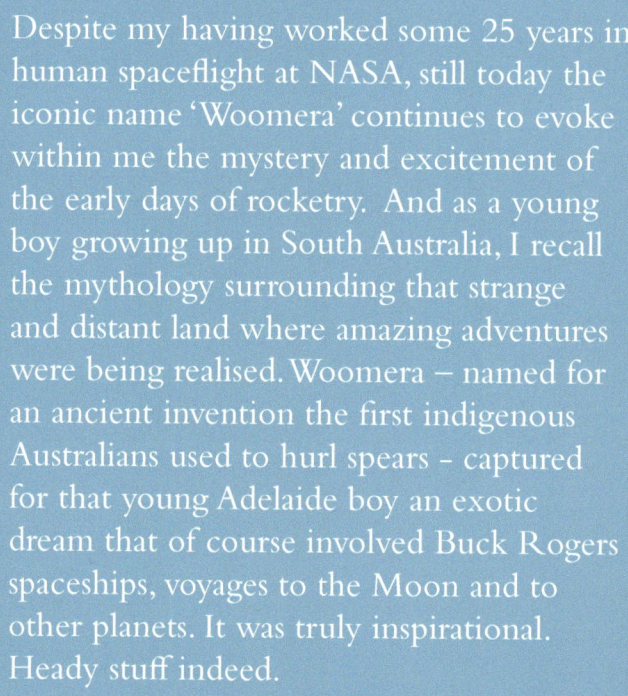

FOREWORD

Despite my having worked some 25 years in human spaceflight at NASA, still today the iconic name 'Woomera' continues to evoke within me the mystery and excitement of the early days of rocketry. And as a young boy growing up in South Australia, I recall the mythology surrounding that strange and distant land where amazing adventures were being realised. Woomera – named for an ancient invention the first indigenous Australians used to hurl spears - captured for that young Adelaide boy an exotic dream that of course involved Buck Rogers spaceships, voyages to the Moon and to other planets. It was truly inspirational. Heady stuff indeed.

But even if that young boy's imagination embellished the reality, Woomera did pioneer significant developments in launch vehicle testing and ultimately allowed Australia to help open the doorway to space by becoming, in 1967, one of the first nations to build and launch its own satellite. Although the efforts behind that milestone did not evolve further, the Woomera site has continued to host sounding rocket and missile tests, and has seen re-entry vehicle testing, hypersonic and SCRAMJET testing. Recently it was the re-entry and landing site for the Japanese Hayabusa probe that returned micro-samples of asteroidal material.

LEFT *Australia from space. Looking back toward Earth from the Moon during the Apollo 11 mission in July 1969. The unique red of the Australian outback can be seen on the left of the globe, partially covered by cloud. (Photo courtesy of NASA)*

But it is not just the Woomera site where Australia has contributed and participated in space exploration on the international stage. Australia's antipodean location offered a unique geographic opportunity to host the early tacking sites for the Mercury, Gemini and Apollo missions. Indeed, the first images and words from the Apollo 11 crew on the surface of the Moon came by way of the tracking stations in Australia. And I recall, like anyone growing up in Australia in the sixties, the great national pride we felt when the city of Perth turned on its lights for John Glenn's historic orbital flight. Little did I know that thirty-five years later, I would have the opportunity to actually meet and talk to John Glenn about that legendary moment. But even more surprising was that my own home town, the city of Adelaide, would come to do the very same for me as I first orbited the Earth in the space shuttle Endeavour in 1996.

The story behind these and many other developments are recorded in Australia in Space. And as you read this history of the Australian role in opening the space frontier, it is my hope you will appreciate how these events came about, and understand the efforts of the dedicated individuals who were behind them. You, too, may be surprised as you share in the excitement of what was done by those engineers and scientists who were motivated by a sense of national pride and curiosity. They knew they were participating in the start of a great adventure.

Below Andy Thomas proudly displays an Australian flag on board the Space Shuttle Endeavour during STS-77, his first spaceflight. (Photo courtesy of NASA)

Andrew Thomas
NASA Astronaut (Retired)
Houston, Texas

Above (top) Members of the Australian Rocket Society (ARS) pose with three different rockets prior to another failed launch attempt in March 1937. (Philatelic Collection, Martin Walker)

Above (bottom) President of the Australian Rocket Society, Alan Young, with the society's first purpose-built mail rocket, Zodiac. (Photo courtesy of Status International Sydney)

CHAPTER 1

Countdown: Australia's First Rocketeers

Rocketry is the basic technology needed for spaceflight, since the rocket is the only form of propulsion that can operate in the vacuum of space. Rockets function according to the 'Reaction Principle', which is described in Sir Isaac Newton's Third Law of Motion: 'For every action there is an equal and opposite reaction.' Therefore, unlike propellers, jet engines or balloons, rockets do not need to be surrounded by an atmosphere in order to work, since they carry their fuel, and the oxygen needed to burn it in order to generate thrust, on board.

While Australia cannot claim any significant contributions to the early development of rocketry, local inventors, dreamers and entrepreneurs were inspired by overseas developments to undertake their own rocketry experiments.

BACKGROUND IMAGE THIS PAGE (TOP) Postcard carried aboard the Australian Rocket Society's unsuccessful flight celebrating the silver jubilee of King George V in October 1935. (Photo courtesy of Status International Sydney)

LEFT The decorative sticker, or vignette, used on ARS rocket mail. It shows stylised designs of the ARS' first two rockets, the Zodiac and the Orion, which actually looked quite different. (Courtesy of Status International Sydney)

A nineteenth century Rocket Experimenter

By the middle of the nineteenth century, gunpowder-powered rockets, based on a technology first developed in China about 1,000 years ago, were being used in both civilian and military roles. Military rockets, such as the British Congreve rocket, first demonstrated in 1805, were used as a type of artillery, as incendiaries and for signalling. Civilian uses for rockets included not only celebratory fireworks and maritime signalling, but also rocket-propelled harpoons for whaling and maritime rescue rockets, for sending lifelines to stricken ships as an aid in rescuing the passengers and crew. One of the first successful rescue rockets was developed on Britain's Isle of Wight by John Dennett. In 1826, he developed a modified Congreve rocket that could be fired from the shore to a stranded ship, carrying a lifeline that could then be used to get people to safety.

'Dennett's rockets', as they were known, began to be used in Australia around the 1850s. They were supplied to rescue stations around the country's southern coasts, situated in areas prone to shipwreck. However, because they were sent by sea from Britain to Australia, and might then be in storage for some time before being used, many of the rockets failed, due to the deterioration of their gunpowder charges. Because the failure of the rescue rockets endangered lives, a South Australian chemist, Frederick Hustler, was inspired to become Australia's earliest known rocket experimenter.

An industrial chemist from Port Adelaide, Hustler's hobby was pyrotechnics; by the 1870s, he had turned manufacturing fireworks and signal lights into a profitable side business. Disturbed by the frequent reports of Dennett's rocket failures, Hustler set out to develop an improved rescue rocket that, by being locally manufactured, would be less prone to failure than the imported ones. Hustler's new rocket was first publicly demonstrated on 20 June 1873, to Accession Day holiday crowds at Adelaide's Semaphore Beach. Newspaper reports of the event compared its performance favourably with that of Dennett's rockets.

Hustler tried to interest the South Australian Colonial Government in purchasing his new rockets to replace the unreliable Dennett's and a formal comparison trial took place in September 1874. However, although Hustler's life-saving rocket proved itself to be superior to the imported model, the South Australian authorities declined to place an order for them and the project was abandoned. As this book will show, Hustler's disappointing experience could be seen as a foretaste of a situation well known to the Australian space community in the twentieth century, with

Above (top) An early illustration of a Dennett's rocket used for a marine rescue. (From the author's collection)

Above (bottom) Although a later development, this Boxer rescue rocket, used by the Swansea, NSW, Rocket Brigade, is very similar in appearance to a Dennett's rocket. Hustler's rocket also probably looked very similar to this. (From the collections of the State Library of New South Wales)

promising innovations and the development of local space expertise unable to be capitalised upon, due to lack of government investment.

The Rocketeers of the Spaceflight Movement

Australia's next rocket experimenters did not appear until the 1930s, when a handful of local 'rocketeers' were inspired by the international Spaceflight Movement. The Spaceflight Movement, which began in the latter half of the 1920s, grew out of the work of two great pioneers of spaceflight: Konstantin E Tsiolkovski, a Russian theoretician who established all the basic mathematical laws of spaceflight and published the first major work on astronautics in 1903; and Robert H Goddard, an American physicist, who designed and flew the world's first successful liquid-fuel rocket in March 1926. Their work, combined with the rise in popularity of science fiction (the first American science fiction magazine, Amazing Stories, was also published in 1926), encouraged the establishment of space travel and rocket societies in many countries, which undertook theoretical and practical research in rocketry and spaceflight.

The first known Australian rocketeer of the Spaceflight Movement was Brian Falkenberg, who lived on a sheep-farming property outside the small rural town of Byaduk in the Western District of Victoria. A science fiction fan and amateur astronomer, Brian seems to have gained at least part of his knowledge of rocket science from the pages of Amazing Stories and other science fiction pulp magazines of the period. In 1931, seventeen-year-old Brian designed his own rocket, an unusual contraption 10 inches (25.4cm) long and powered by what he described as six 'flame rocket tubes' that may have been fireworks. Although he described in a letter to the American Interplanetary Society, in December 1931, some tests in which the rocket propelled itself across the ground on four wheels, it is not known if Brian ever successfully launched his rocket. Despite his youthful enthusiasm for rocketry, by the mid-1930s Brian had turned his attention to amateur radio, which became his life-long passion. He ceased his rocket experiments, went on to inherit the family property and stepped out of the history of Australian rocketry.

An offshoot of the Spaceflight Movement was rocket mail, which strongly influenced Australia's other rocket experimenters of the 1930s. The founder of rocket mail is generally considered to be Friedrich Schmeidl, an Austrian who wanted to develop a rocket-based postal system, to overcome the difficulties of communication in his country's mountainous terrain. In 1928, Schmeidl began to finance his research by producing and selling souvenir envelopes carried in his experimental rockets, an innovation that quickly grew into a popular and lucrative branch of philatelic collecting. While some mail rocketeers were serious experimenters, others were philatelic groups interested in producing commemorative envelopes that would be flown by adapted firework rockets or maritime rescue rockets.

In 1934, Alan Hunter Young, a Brisbane architect who had already founded the Queensland Air Mail Society, was inspired by Schmeidl and other mail rocketeers to conduct rocket mail flights in Australia. Young organised the flight of the first Australian rocket mail on 5 December 1934, using a standard ship's rocket to carry the philatelic payload, stored in a metal container attached to the rocket by a long life-line. This 'mail rocket' was launched from a ship anchored in the Brisbane River near Pinkenba, to the north bank of the river. Following a second rocket mail flight in August 1935, in which two ship's rockets carrying philatelic items were fired between Fraser Island, off the coast of Queensland, and the offshore wreck of the SS Maheno, Young and his friend Noel Morrison founded the Australian Rocket Society (ARS), Australia's first rocketry group, in October that year.

While clearly interested in the philatelic side of rocket mail, Young viewed himself as a serious rocket experimenter, with an ultimate aim similar to that of Schmeidl: to develop a rocket-powered postal delivery system, which would carry mail to the remotest areas of Queensland faster than a train or aircraft, or deliver mail and packages from ships to islands. Although Young obtained the technical details of overseas mail rockets, it appears that he and Morrrison were not highly technically skilled themselves, and although Young claimed the title of its 'designer', the Australian Rocket Society's

Above Postcard carried aboard the Australian Rocket Society's unsuccessful flight, celebrating the silver jubilee of King George V in October 1935. It was re-flown on the ARS' next attempted rocket firing, which also failed, dumping its cargo of mail into the river. (Courtesy of Status International Sydney)

first mail rocket, Zodiac, and its successors were constructed by a Brisbane plumbing firm. The Zodiac, its sister rocket, Orion, and the ARS' Test rocket series were all powered by home-made gunpowder charges, devised by Alan Young.

Between October 1935 and May 1937, the ARS launched eight experimental rockets, the majority of which were unsuccessful due to failures or malfunctions of their home-made gunpowder propulsion system. The Australian Rocket Society's first successful flight finally occurred in July 1936, when R.T.6 (the '6' stemming from the fact that it was the sixth rocket mail flight Young had attempted), successfully flew across the Brisbane River at Moggill Ferry. Smaller than the ARS' earlier unsuccessful rockets, R.T.6 was about one metre in length. It travelled 350 feet (107m), reaching an altitude of 400 feet (122m) before it nose-dived into the ground on the opposite bank of the river.

This success was repeated with the following flight, R.T.7. Under the supervision of the police, this rocket was launched at Brisbane's Enoggera Rifle Range on 24 September 1936, as an event in conjunction with the Brisbane Philatelic Exhibition. A tiny rocket, R.T.7 was approximately 2.5 feet (0.76m) in length and weighed only 7.5 pounds (3.4kg). Launched from a simple guide-rail, it reached a maximum height of 200 feet (61m) and travelled some 300 yards (274m), landing within five yards (4.6m) of a target mark.

Unfortunately for the ARS, R.T.7 was its last successful rocket flight and, after further failures of ambitious launch attempts, the society's final launch took place in May 1937. Apparently stung by criticism that they were only carrying out stunts in order to raise money from philatelic sales, and having exhausted their financial resources, Young and Morrison decided to cease their rocket experiments and the Australian Rocket Society was defunct by the end of 1937.

Australia's last Spaceflight Movement rocketeer, and its youngest, was James Kenneth Atock, known as Ken, a Victorian high school student with a passion for space travel and an interest in rocket mail. Academically gifted, while studying at Melbourne's prestigious Camberwell Grammar School, Ken published his first professional newspaper articles about rockets and space exploration in 1936, at just 15 years of age. Through correspondence with the Australian Rocket Society and overseas mail rocketeers, Ken gained information on their rocket designs. He designed and built his own mail rocket, which he attempted, unsuccessfully, to launch on 6 October 1936, shortly after the Queensland ARS' successful R.T.7 rocket flight. The flight attempt was conducted at Fisherman's Bend, an industrial area of Melbourne that would, co-incidentally, later become associated with the Australian Defence Scientific Service, which undertook some of Australia's earliest space-related defence research.

Ken's rocket flight was intended as a 'demonstration' launch to raise interest in the idea of a special rocket mail flight in conjunction with the Australian Air Mail Exhibition, to be held in Melbourne in 1937. Although this first launch attempt was unsuccessful, with the rocket catching fire before it even left the ground, Ken went on to construct a new rocket, called Mercury, which was intended to be fired from Lilydale, Victoria, carrying 500 mail items, as a celebration of the coronation of King George VI on 11 May 1937. This launch was, however, cancelled at the last moment by the State Explosives Department, on the grounds of public safety. Although an attempt was made to find an acceptable alternative launch site, either further inland, or even from a platform at sea, the Mercury was never to fly, although it was displayed at the Melbourne Motor Show a few days later.

Ken Atock's potential to become a more important rocket experimenter was never to be realised. In 1939, shortly after the outbreak of World War II, Ken enlisted for service in the army, falsely giving his age as 20, since at 18 he was below the age limit for front-line soldiers at the time. After being interned in a POW camp on Crete, he was killed in July 1941, while attempting to escape in order to deliver intelligence information to the Allies. In 1976, the 'Kenneth Atock Memorial Scholarship' was established at Camberwell Grammar, to support students' academic performance in science with the emphasis on space and rocketry. This scholarship is still awarded today.

Although early rocket experimentation ceased in Australia in 1937, and left no technological legacy on which future Australian space projects could build, Australian science fiction authors during World War II began to envisage future space missions being launched from 'down under'. The establishment of the Woomera Rocket Range just a few years after the war would provide a real-life launch facility from which new adventures into space could blast off...

RIGHT *Ken Atock, just fifteen years old, and his school friend Bob Ware, unloading and inspecting the mail payload from his unsuccessful first rocket. (Courtesy of W. Holmes)*

OPPOSITE BELOW *The Australian Rocket Society's second successful rocket, R.T.7, was flown at Brisbane's Enoggera Rifle Range in September 1936. Young is shown holding the rocket, at its target mark, in company with the police supervisor Sergeant Edward Creedy of Newmarket Station. (From the collection of the Queensland Police Museum. PM0723)*

Professor Frank Cotton

Great-grandfather of the Spacesuit

Modern American spacesuits trace their lineage back to the first partial pressure suit for high altitude use, developed at the end of World War II by Dr James Paget Henry. Henry developed his design by adapting the pneumatic anti-g suits worn by US military pilots. What he may not have known, however, is that the American pneumatic anti-g suit itself owed a legacy to the work of Australian medical researcher Professor Frank Cotton.

Professor Francis (Frank) Stanley Cotton, of the University of Sydney, is best known for his pioneering post-war research in sports medicine. However, during the war he turned his attention to aero-medical work, believing that his research interests in blood circulation and the physiology of exercise might provide the answer to the problem of 'blackout' – a loss of vision, often followed by loss of consciousness, when pilots made high-speed turns or steep dives and pullouts. This condition was caused by the acceleration forces created during aerial manoeuvres, which drained blood away from the head into the lower limbs, making it difficult for the heart to pump blood to the brain, which then suffered from a lack of oxygen.

Cotton believed that the best way to counter blackout was to maintain an adequate supply of blood to the head: if pressure could be applied to prevent blood from pooling in the lower part of the body and force it back up to the heart, then the heart itself would have the strength to maintain the blood supply to the head. He conceived the idea of an air-filled anti-g suit that would apply pressure to the pilot's lower body to force the blood back up into the torso.

Working independently of overseas researchers, by the end of 1941 Cotton developed a 'graded pressure' garment, with the assistance of the Royal Australian Airforce's (RAAF) Flying Personnel Research Committee and Dunlop Rubber Australia. His 'pneumatic anti-g suit' design consisted of a pair of shorts (trunks) and two separate, booted leggings, which could be adjusted to fit pilots of various sizes through the use of zips and lacings. The suit was constructed from two layers of rubber latex, with the rubber air-sacs sandwiched between. The outer layer was reinforced with parachute silk, which prevented it from expanding, but allowed for ease of movement.

BELOW *Professor Frank Cotton and his team pose for a group photo shortly after the first flight trial of the anti-g suit at Mascot Airport, Sydney, in 1941. Cotton is standing to the right of the test pilot. (From the author's collection)*

RIGHT Cotton's research assistant, Ken Smith, about to enter the Sydney University centrifuge for a test of the anti-g suit prototype. Limited funding meant that Cotton used jury-rigged and second-hand materials to construct his centrifuge. (From the author's collection)

When the air sacs were inflated, the inner layer expanded against the pilot's skin, producing a pressure closely proportional to the g-force applied to the body.

Both the shorts and the leggings contained air sacs that together provided a six-step pressure gradient along the body, from the feet to the base of the ribs. The greatest pressure was applied around the ankles, with the other pressures progressively diminishing toward the heart. Carbon dioxide was used to inflate the pressure bladders, and the suit was connected by tubing to a cylinder of the gas. The pressure in the bladders was controlled by the centrifugal force of the aircraft acting upon a unique 'column of pistons' valve designed by Cotton, who was always an innovator in experimental techniques.

To assist his research, Cotton arranged for the construction of a human centrifuge, only the second one in use among the Allies at the time. This centrifuge, housed in a room at Sydney University, had a unique design that was a wonderful example of Australian 'make-do' innovation in the face of wartime materiel shortages. With its canopy of Masonite, the centrifuge turntable revolved on wheelbarrow wheels, driven by two secondhand 10 hp (7.5 kW) motors. This enabled the centrifuge to achieve a constant angular velocity of 60 rpm, producing a maximum acceleration of about 9.5 g at a radius of 9 feet (2.8m). Even the accelerometer was a uniquely jury-rigged apparatus, as no suitable accelerometers were available in Australia at the time.

In late November 1941, only a few weeks before the outbreak of war in the Pacific, Cotton travelled to Canada, the United States and Britain, to confer with blackout researchers and disseminate his work. Cotton found that his blackout research was well in advance of that being done in the United States, and his air-filled suit design was adopted and adapted by American researchers, underlying the design of the first successful US anti-g suit, on which Dr Henry would later base his partial pressure suit.

The production version of Cotton's invention, known as the 'Cotton Aero-dynamic Anti-G' (CAAG) suit was first deployed operationally in Australia in 1943 and by the end of that year, a one-piece version of the suit was developed. The CAAG suit was, however, never used in actual combat. After the war, the RAAF decided that it would not continue with the development of the CAAG suit, but would purchase future anti-g suits from the United States, a precursor of a situation that would recur repeatedly in the story of Australian space activities.

Although it is not certain that Professor Cotton had any specific idea of the application of his anti-g suit research to the development of a pressure suit, he must have had some idea of the link between the two and the potential for the ultimate development of a space suit, for, in 1942, he made the following prophetic remark to a research assistant: 'As a result of my work, Man's now going to go to the Moon'!

Above Computer Judith Ellis manually recording data from a film record in 1949. She is using an overhead projector to examine an individual frame of film from a tracking instrument. (Photo courtesy of the Defence Science and Technology Group)

Right Human computers and electronic computers worked together on processing the huge quantities of data from the weapons trials at Woomera. (Photo courtesy of the Defence Science and Technology Group)

CHAPTER 2

Ignition! The Origin of Space Activities in Australia

Australia's involvement with space activities commenced with the beginnings of the Space Age in the latter half of the 1950s. The nation's first space projects were centred on the Woomera Rocket Range, a vast outdoor weapons testing facility that had been established in Australia by the United Kingdom shortly after World War II. Although space travel was still the stuff of science fiction at the time the Range was set up, the specialised skills in missile and tracking technologies developed there to support its operations would make Woomera the natural home for Australia's earliest space activities.

BACKGROUND IMAGE THIS PAGE *The first sounding rocket launched at Woomera was the Skylark, a long-lived research rocket originally developed as part of Britain's contribution to the International Geophysical Year. (Photo courtesy of the Defence Science and Technology Group)*

Rockets in the Desert

Why did Britain want to establish a missile test range in Australia? During World War II, German scientists and engineers developed the world's first long-range missile. Dubbed the Vergeltungswaffe 2 (Vengeance Weapon 2 or V2), this terror weapon was used against Britain and other European countries in the last year of the war.

The V2 demonstrated that long-range missiles had the potential to be effective strategic weapons, capable of striking at great distances with a powerful warhead. Recognising their strategic value, the major powers began to establish rocket and missile programs as the war in Europe came to an end. The United Kingdom was one of many countries that decided to develop a missile capability, commencing a significant program of research and development in the field of guided weapons.

To accommodate the missiles that it was planning to build, Britain needed a test range up to 1600 kilometres long, with regular good visibility to allow visual tracking of the missiles during test flights. It was also preferable to have the whole flight take place over land, so that the missiles could be recovered in the event of malfunctions for assessment of developmental problems. These conditions could not be found in the UK, a small and densely populated country, surrounded by busy shipping lanes, but they could be found in Australia.

Australia's size, its cultural, social and political ties to Britain, and its economic and political stability, made it highly suitable as a location for the type of weapons testing range the British envisioned. With most of its population concentrated in the south-eastern coastal regions, a vast, sparsely-inhabited tract of territory across central Australia could be declared a restricted security zone, to reduce

BELOW A captured V2 rocket on display at an airshow at the former Mallala Airport, near Adelaide, in 1954. The V2 was originally brought to Australia for training purposes in connection with the early plans for missile testing at Woomera. (Photo courtesy of Dr Brett Gooden)

the presumed risk of Communist espionage and sabotage, with little overall impact on the country. The harsh but fragile environment of central Australia was deemed far less important than its remoteness, which was considered an important security asset in the developing Cold War. Thus Britain's need for vast areas in which to test captured V2s and their technological descendants would eventually carry Australia into the forefront of early space activity.

Why would Australia allow its territory to be used for British weapons tests? The Rocket Range proposal was a component of Britain's policy of 'Empire defence', under which its weapons development, testing and manufacturing programs would be dispersed around its Empire (actually in the process of transitioning to the modern British Commonwealth), to prevent their destruction in any future attack on the United Kingdom. Australia potentially stood to receive a considerable portion of this work, being far removed from the expected arenas of future conflict. With the Cold War looming, the Labor Government of Prime Minister Ben Chifley (1945-49) saw participation in the Empire defence program as an opportunity to give loyal support to the former colonial motherland, while at the same time advancing Australia's own interests.

World War II had forcibly brought home Australia's vulnerability, due to the lack of a solid industrial manufacturing base capable of producing weapons, aircraft and other military technologies. The Australian Government had bitter memories of being denied armaments from Britain, for use in the Pacific War, on the basis that they were needed in Europe. Chifley, therefore, saw the proposed weapons testing facility as a means for Australia to ensure access to the latest weaponry in the event of another war. Although Australia's industrial plants and engineering skills had advanced rapidly during the war, many of those plants became idle and skilled workers unemployed once the hostilities ceased. Chifley envisaged potential benefits for the development of Australian technology and post-war employment through the establishment of a British armaments industry in the country.

Consequently, the Chifley Government agreed to the weapons facility proposal with the one firm proviso that Australia was to be an equal partner in the project. Australia would share the development and maintenance costs of the Rocket Range in return for technology transfer, the employment of Australians within the facility and contracts to Australian industry. From this agreement, the Anglo-Australian Joint Project was born, a co-operative weapons development project that was to last until 1980 and pave the way for Australian involvement in space.

Following the Chifley Government's approval of the Rocket Range proposal, a suitable site had to be found for the establishment of the rangehead and its supporting facilities, which would include a town in which the Range personnel and their families would live. An area near Lake Torrens, about 480 kilometres north-west of Adelaide, was chosen as the location of the new testing facility. Subsequently, both the Range and its associated township were given the name 'Woomera', from an Aboriginal name for the throwing stick device used by the Indigenous people to give increased range to spears. From the Woomera Rocket Range, missiles could be launched and tracked and bombs and other weapons could be tested, with any test failures presenting minimal danger to life and property.

Two vast areas were declared as 'prohibited areas', to be developed for the Rocket Range: a short-range impact area in South Australia, extending about 400 kilometres from the rangehead, and the Talgarno impact area in north-western Western Australia, some 1800 kilometres downrange from the launch area. An impact area in the vicinity of Christmas Island, in the Indian Ocean, about 2400 kilometres from Australia's north-east coast, was also planned for the testing of the longest range missiles then envisaged. The Long-Range Weapons Establishment (LRWE), which was to oversee the administration of Anglo-Australian Joint Project and the operations of the Woomera Range, established its headquarters in a disused munitions factory at Salisbury, near Adelaide. This site would become the backbone of Australia's defence science establishment.

Woomera Rocket Range and the Weapons Research Establishment

In 1947, the LRWE began operations, with the object of building the weapons test range starting from Woomera and extending in a 400km-wide swath across central Australia. Because it was so remote, Woomera was an extremely expensive research centre to establish and operate: the cost of maintaining each person and each facility there was two-and-a-half times what it would have been if the Range had been established close to a capital city. At the height of activity at Woomera, the township boasted a population of around 6200 scientists, engineers, technicians, support staff and their families, who had to be supplied with all the necessities of everyday life, in an arid area with no permanent water supply.

Although the Joint Project was almost abandoned in 1948, when Britain decided to cancel its long-range missile development program, Australia, having already invested considerable financial resources in the establishment of the Range, mounted a campaign to keep the project alive. As a result, a small general purpose range was constructed, extending only 48 kilometres, to test a variety of defensive weapons. In 1949, it was decided to develop a broader background for the work on guided weapons, and to cover other aspects of defence science such as radar, bombs and other armaments, and aircraft. This new focus rapidly made it clear that Australia lacked capability in many of the scientific and technical fields that were required to support this research. Consequently, 1949 also saw the establishment of

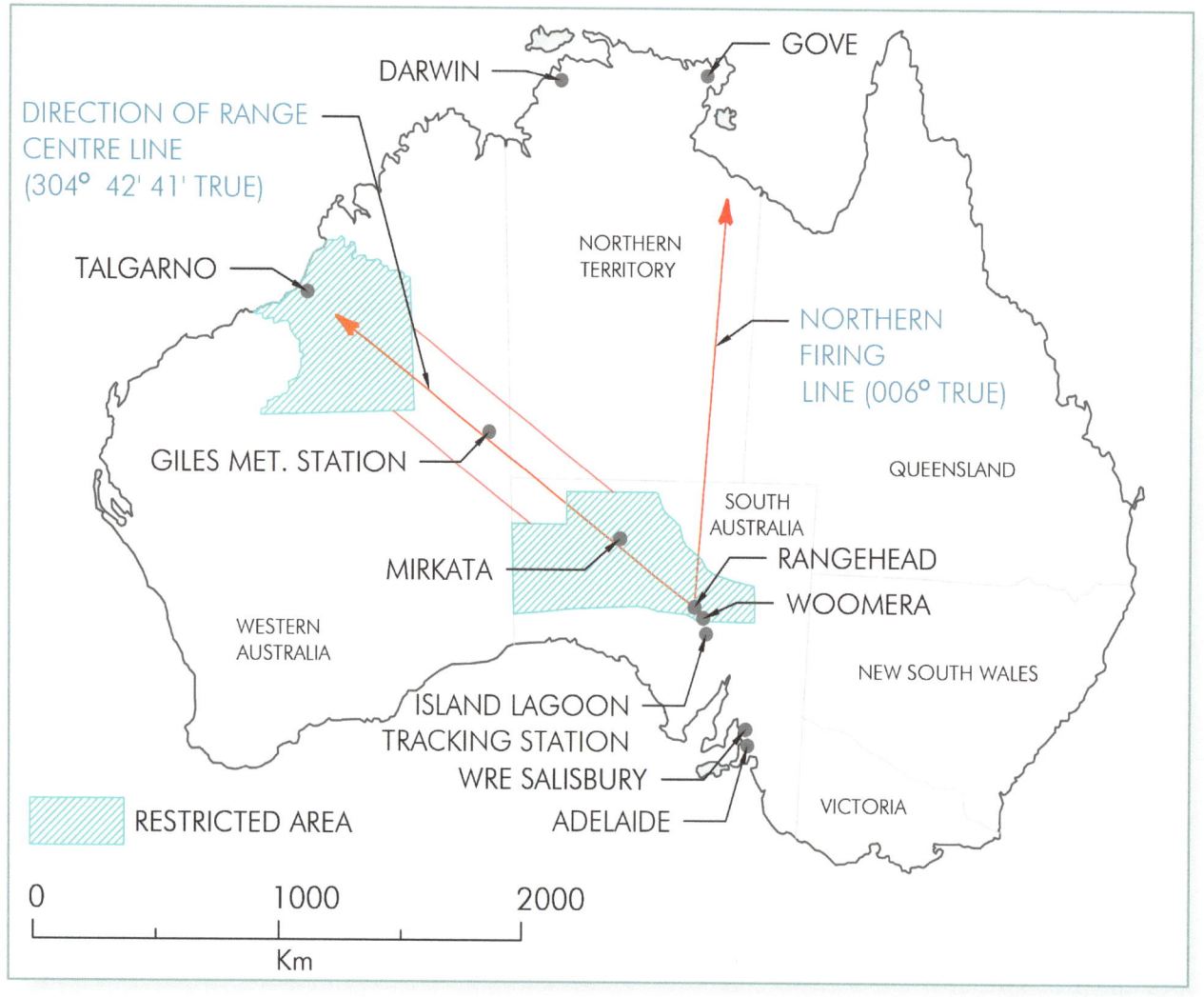

Above This map shows the location of the Woomera Rocket Range prohibited areas in South and Western Australia, and the locations of various facilities associated with the Joint Project and other space operations centred on Woomera. Long-range missiles were originally planned to be fired along the Range centre line, while later satellite launch attempts were fired in a northerly direction.

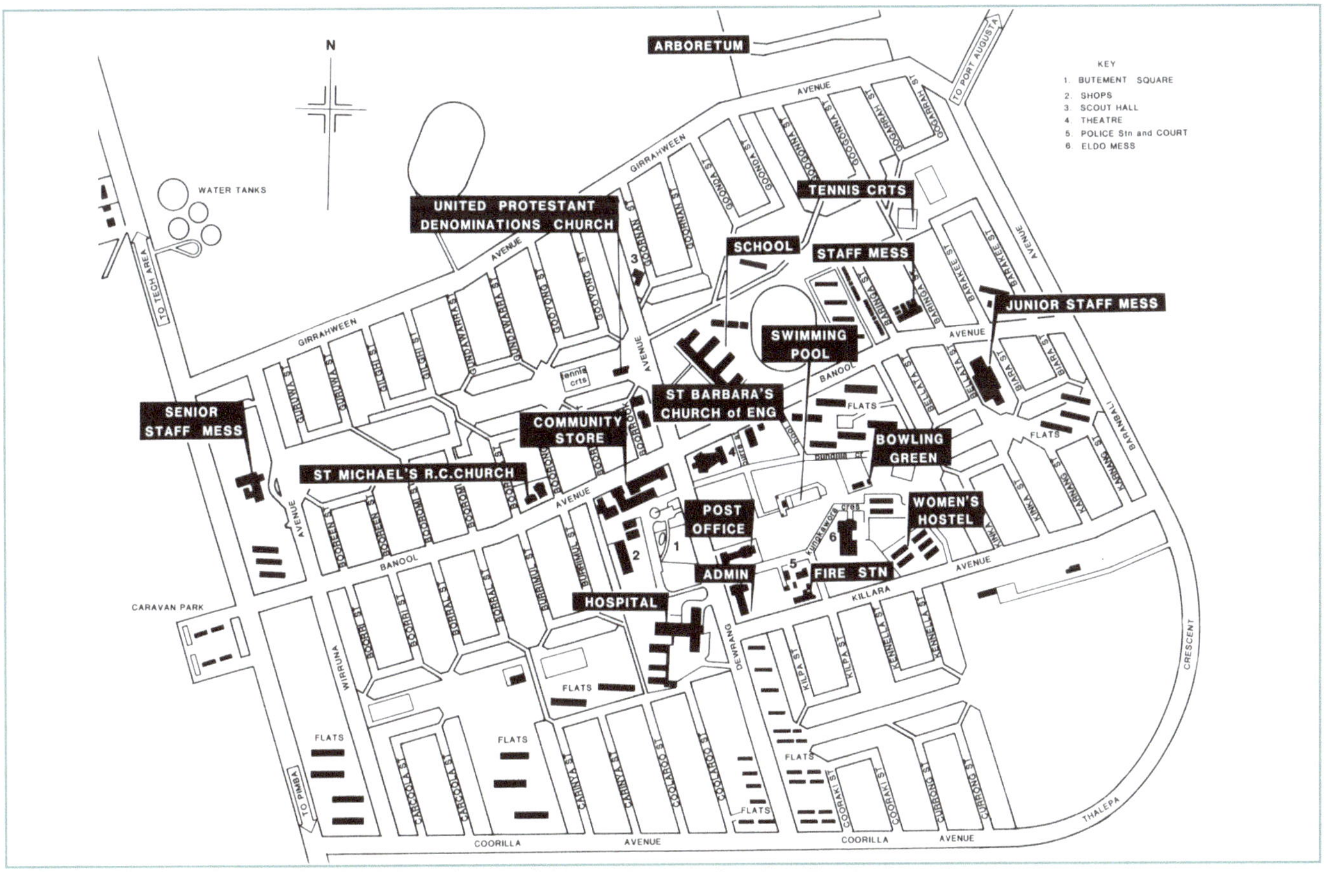

THIS PAGE *In order to attract staff to work in such an isolated location, Woomera had to provide every amenity possible. (ABOVE) Aerial view of the township in 1962, showing some of the town's churches and recreational facilities. (BELOW) A plan of Woomera in 1965, showing the compact layout of the township and the location of the major amenities. (Images courtesy of Defence Science and Technology Group)*

the Australian Defence Scientific Service (ADSS), to consolidate and expand the nation's defence-related research and development efforts.

This new agency encompassed the Long-Range Weapons Establishment and the existing Defence Research Laboratories. In 1951, the ADSS established three new research laboratories at Salisbury — in high-speed aerodynamics, propulsion, and electronics research — to provide research and development support for specific Australian weapons projects at Woomera. In 1955, the LRWE and the Defence Research Laboratories were combined into a new organisation, the Weapons Research Establishment (WRE), which remained the major component of the ADSS until the Service was disbanded in 1974.

With the consolidation of the WRE, the objective of the Joint Project was again redefined. Britain and Australia would now collaborate on the establishment and operation of an experimental range and supporting facility for the testing and development of long-range guided weapons, pilotless target aircraft and air-launched equipment, including radio and radar control, and other projects that could be carried out making use of facilities then in existence or planned. Australia would provide range and support facilities, developmental facilities and production capacity. Much of the equipment for the Joint Project at Woomera was supplied by the WRE itself, with significant participation by Australian and overseas industry.

Operationally, the Woomera Rocket Range actually consisted of a number of separate test ranges, which were used for varying purposes, from bomb trials to rocket launches. The principal missile range was based at Koolymilka, about 40 kilometres north-west of the Woomera township, where several separate launch aprons allowed work to proceed on multiple projects simultaneously. Adjacent to the main rangehead were airstrips and facilities for the operation of pilotless target-aircraft, such as the Australian-designed and built Jindivik, while an elaborate network of electronic instrumentation, radar and tracking cameras covered the trials areas. By 1957, Woomera was considered to be the best equipped and managed test range in the Western world, able to conduct a major program with fewer staff than at overseas facilities. As later chapters of this book will describe, during its peak years in the 1960s, Woomera was the most heavily used space launching facility in the world, apart from Cape Canaveral in the United States.

Despite popular misconception, the infamous British nuclear tests that were conducted in Australia between 1953 and 1963 were not directly connected to the activities at Woomera. Although there were unfounded early rumours that flying atomic bombs were going to be launched there regularly, devastating large areas of the outback in the process, only inert ballistic dummies of nuclear bombs were ever dropped on the Woomera Range during bombing trails.

Although the first British nuclear test site on the Australian mainland, Emu Field, was carved out of a section of the Woomera Range, it was considered too remote for regular use and only two atomic devices were tested there. The remaining nuclear tests were conducted at the Maralinga Prohibited Area, which was adjacent to the Woomera Range, but separate from it. These tests were carried out under a separate agreement with Britain that was not connected with the Joint Project. This agreement was established in 1952, under the Menzies Liberal Government, which may have agreed in part because it wished ultimately to obtain nuclear weapons for Australia's defence.

Unlike the activities at Woomera, Australia had little control over the conduct of the nuclear tests, which have left large areas of radioactively contaminated land in central Australia and created long-term health problems for Australian military personnel involved with the tests as well as the Aboriginal inhabitants of the area. The WRE's involvement with the atomic tests was confined to providing logistic support for the nuclear programs. Woomera also functioned as a recreation centre for personnel from the test sites, which were supplied by the RAAF from its base at Woomera.

Early Space-related Research and Development at the ADSS

Although the major weapons and missile projects to be conducted at Woomera were British, Australia intended to establish its own weapons development and testing programs on a smaller scale. Since much of the ADSS' research was focused toward missile development, its scientists and engineers produced innovations that either contributed to Australia's early space activities or could have formed the basis of a more extensive national space effort in the 1960s, had the Australian Government decided to establish such a program.

Due to the advanced nature of much of the work being conducted at Woomera, the scientists, engineers and technicians of the WRE often had to build the technology they needed to their own designs and specifications, which allowed them to develop considerable expertise in many areas of advanced space-relevant technology. Some of their most important innovations include:

Computing, Automated Data Processing and Mathematical Systems Modelling

By the end of the 1950s, the WRE's Mathematical Services Group (MSG) had become a world leader in the development of automated data processing, mathematical modelling and computer simulation of missile systems.

Processing the data recorded for the test activity at Woomera was initially a time-consuming task, because all the early monitoring instruments on the Range recorded their data on film. The labour-intensive task of reading the film by eye to retrieve the data, and then calculating the results, was carried out by 'human computers', who were mostly young women with some mathematical training, their only aid a mechanical calculator. The sheer length of time required to process even the simplest trials spurred the LRWE to begin investigations into the use of electronic computing and data processing technologies.

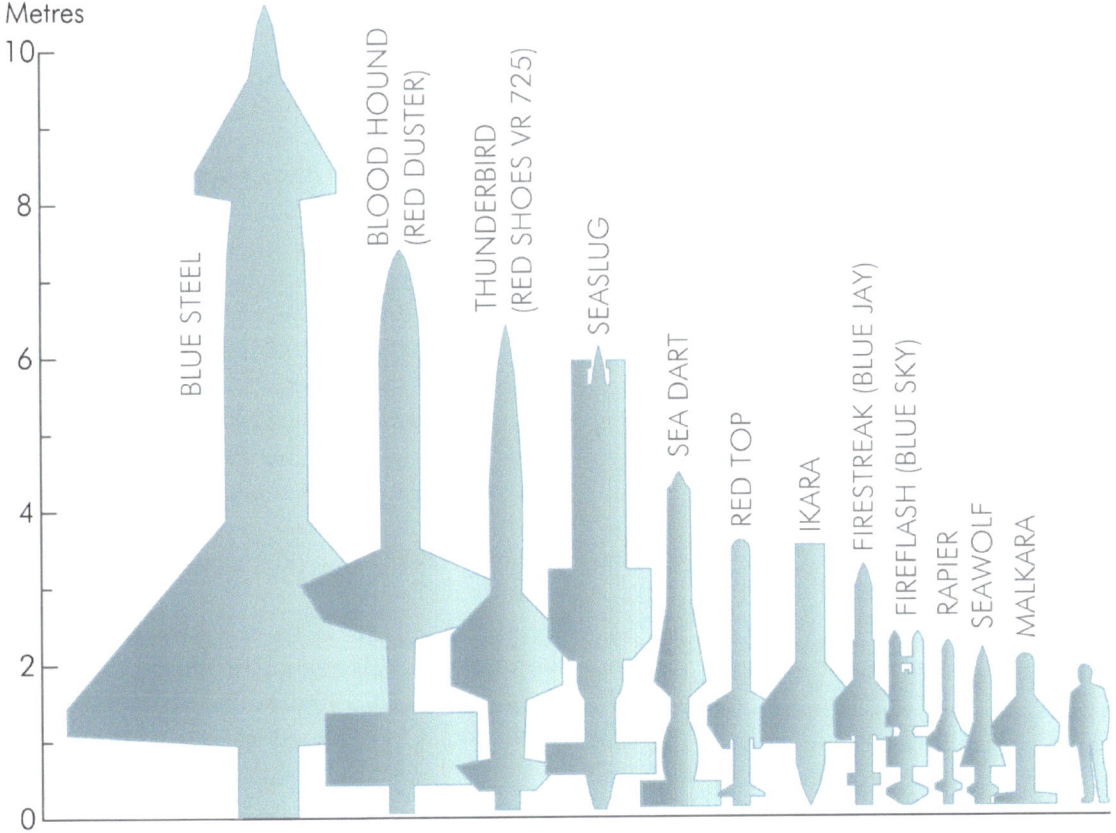

Above Diagram of the major guided weapons tested at Woomera. All were British except for the Malkara and Ikara, which were Australian, and powered by rocket motors developed by the Australian Defence Scientific Service (ADSS). The Malkara and Bloodhound missiles were the first developed using mathematical modelling and computer simulation pioneered by the ADSS. (Based on an original 1960s WRE diagram)

In 1950, the MSG began to use IBM Hollerith punched-card machines for the tabulation of trials data and soon designed a range of modified Hollerith equipment to automate the computing and tabulation of trajectories derived from several types of Range instrumentation, after the film had been manually read and typed onto punch-cards. By 1953, it was clear that hundreds of human computers would be needed to cope with processing the vast mass of data expected to be produced by the missile trials planned for the rest of the decade. This made the acquisition of an electronic computer a cost-effective option and the WRE decided to purchase a digital computer that would become known as WREDAC (Weapons Research Establishment Digital Automatic Computer). It would be the second of four digital computers operating in Australia by the end of 1956.

Based on a British design heavily modified to WRE-developed specifications, WREDAC's software was written in-house and enabled the production of performance reports from test data provided by WRE-developed analogue to digital converters. The computer's output could be presented on Australia's first line printer and the world's first digital plotters. By 1957, the WRE had developed computer programs for automatically processing kinetheodolite tracking camera data directly to cards, while radar and telemetry instrumentation had been fitted with data converters linked to WREDAC. At that time, it could calculate 1000 trajectory points per day, representing a 90 percent reduction in manual computing time and a 20-fold reduction in computing costs.

The MSG also developed a digital to analogue converter, enabling the WRE's much faster analogue computer, AGWAC (to be discussed below) to share the digital instrumentation data. Working in tandem, the two computers formed a powerful team that could process data from a missile test in almost real time. By the late 1950s, Australian developments in computing and automated trials data processing were not only ahead of Britain, but had achieved capabilities that were still under development at Cape Canaveral and other Western missile test facilities.

Alongside its computing and automated data processing capabilities, the ADSS was an early pioneer of mathematical systems modelling and computer simulation. Testing was expensive and there were considerable cost savings in accurately simulating the performance of a missile or other weapon. The immense quantity of data collected by the tracking and monitoring instruments that recorded each test enabled the development of a statistical database for such quantities as trajectory, attitude, and the internal and external behaviour of the weapon under test. This could provide the basis for developing mathematical models of weapon behaviour that would enable them to be computer simulated under various performance conditions without the expense of an actual flight.

Despite scepticism both in Australia and the UK, the LRWE developed a basic analogue simulator known as ARTVS (Australian Rocket Test Vehicle Simulator). Constructed from locally available components, including surplus World War II valves, by early 1954 ARTVS had successfully simulated the flight of a beam-riding missile, establishing the validity of information derived from a mathematical model.

Below The Australian Guided Weapons Analogue Computer (AGWAC) installed at WRE Headquarters in Salisbury. Used for both trials data processing and simulation work, AGWAC, like all computers of its day, occupied an entire room. (Photo courtesy of Defence Science and Technology Group)

In 1955, AGWAC (Australian Guided Weapons Analogue Computer) commenced operation. The last major analogue computer put into service in Australia, AGWAC, like WREDAC, was a modified version of a British design developed to Australian specifications. Intended for both trials data processing and simulation use, AGWAC was operated by Australian-developed programming and was able to process data in near real-time. Working on simulation programs with ARTVS, it was estimated that the two computers saved the labour of 150 human computers.

The first weapon to be extensively developed using mathematical modelling and computer simulation was the Australian rocket-powered anti-tank weapon, Malkara. The WRE then proposed that the trials for the British Bloodhound surface-to-air missile be planned so as to validate a mathematical model of the missile, allowing much of its performance to be subsequently obtained from the model rather than from real-life trials. Despite initial scepticism, this program was so successful that, in the beginning of 1957, the mathematical model approach became policy for missile tests at Woomera.

WREDAC's capabilities were quickly superseded by the new generation of transistorised computers, and in 1961 it was replaced by an IBM 7090. Although AGWAC remained in operation for simulation projects until the late 1960s, no further Australian computers were developed by the WRE, as it was considered more cost effective to purchase commercial computing systems from overseas. Had the Australian Government made a decision to develop a space industry and space program in the 1960s, the WRE's expertise in data processing and systems modelling would have made important contributions, especially in launch vehicle development and satellite design.

Robust Miniature Cameras and the Ultra-Wide Angle Lens

In order to capture every possible aspect of a missile trial, the WRE developed miniaturised, highly robust cameras that could be mounted to both the missiles and their target aircraft. From 1954 until 1974, the ADSS' Optical Group at Salisbury developed a series of specialised cameras for aircraft and/or missile use, capable of recording still shots, cine film, and, in the case of the WREROC (WRE Roll Orientation Camera), a special slit camera able to capture a view of the horizon as a missile rolled in flight.

These cameras were housed in compact, incredibly sturdy cases that contained and protected their film supplies from impact: the cameras were recovered after their host missiles had crashed to the ground at the conclusion of a test flight so that their film record could be analysed. To avoid the need for heavy batteries to power them, the cameras were connected to the power supply of their host vehicle. The cameras were manufactured by Fairey Aviation Company of Australasia Pty Ltd, the Australian branch of the British company. Like many other UK firms involved in British weapons development at Woomera, it had set up shop at WRE Salisbury.

BELOW The tiny WRECISS (WRE Camera Interception Single Shot) missile camera carried a single piece of film to record the missile's closest approach to its target. This robust device could be recovered after the test and its film retrieved for processing. (Photos courtesy of Defence Science and Technology Group)

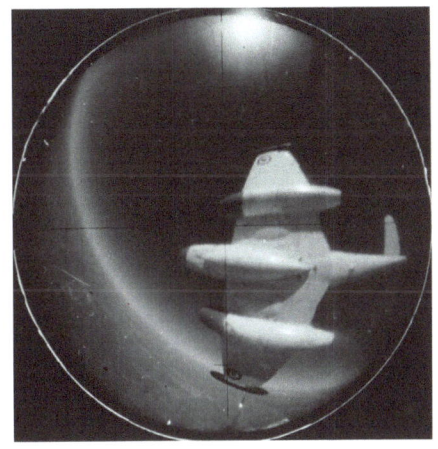

At the heart of each camera was an important Australian innovation, the Ultra-Wide Angle Lens. Developed in 1954 by LRWE Optical Group member Frank Dixon, the Ultra-Wide Angle Lens had an effective 186-degree field of view. Its most important quality for trials work was that its particular shape distorted the image, when recorded on flat film, in such a way that various angles could be measured, which was important in determining particular flight characteristics of a missile. It was also extremely compact and relatively cheap to make. The lenses were manufactured by Australian company Etherington Optical, a specialist optical manufacturer tucked away in the Victorian rural town of Mildura. Although some of the WRE cameras were sold overseas in conjunction with sales of the Australian-developed and built Jindivik target aircraft, and adaptations of the WRE cameras were used in the Australian sounding rocket program, the Australian Government never marketed the WRE camera and lens technology internationally for space use, thus failing to capitalise on a unique Australian product.

Propellant and Rocket Motor Research

In 1949, an early Australian weapons program stimulated ADSS research into solid rocket propellants and led to the creation of the Propulsion Research Laboratory. By 1953, this research had expanded to include studies of the characteristics and chemistry of particular types of solid rocket propellants, as well as investigations into the properties of ablative materials for missile nosecones and high temperature materials used in rocket motor construction, along with rocket nozzle design, materials and manufacture.

By the mid-1950s, the Malkara heavy anti-tank weapon drew on this research, being fitted with the first operational Australian-designed and built solid rocket motor, code-named Marloo. The Marloo was the precursor of a number of ADSS-developed rocket motors, for both weapons applications (such as the Ikara anti-submarine weapon and the Nulka active missile decoy) and almost two decades of Australian sounding rockets, which will be discussed in Chapter 3.

THE INTERNATIONAL GEOPHYSICAL YEAR – A CATALYST FOR SPACE ACTIVITIES

With Woomera Rocket Range already in operation, the impetus for Australian involvement in space activities was provided by the International Geophysical Year (IGY), an international co-operative program for the scientific study of the Earth and its relationship to its surrounding space environment. Undertaken from 1 July 1957 to 31 December 1958, the IGY became a significant catalyst for commencing space-related activities in many countries.

The IGY concept was adopted in 1952 by the International Council of Scientific Unions. In 1955, both the United States and the Soviet Union announced that they would attempt to launch Earth-orbiting satellites during the IGY. The United States named its program Vanguard, and confidently expected to be the first nation to put a satellite into orbit. The Western world was

LEFT *Educational poster produced by the US National Academy of Sciences promoting space research during the IGY. This poster was part of a set of six, each focused on an area of geophysics, created for an IGY outreach project. (Photo courtesy of the National Academy of Sciences, United States)*

therefore stunned when the USSR succeeded in launching the world's first satellite, on 4 October 1957 (actually, the early hours of 5 October Australian time). Named Sputnik 1 (meaning 'travelling companion' in Russian), the Soviet satellite was much larger than America's planned Vanguard. In the Cold War climate of the time, space achievements immediately came to be seen as evidence of technological — and supposedly political and ideological — superiority. This rapidly led to the development of the 'Space Race' between the US and the USSR, with both countries striving to achieve status-conferring space 'firsts'.

Satellite Tracking – Australia in the Right Place at the Right Time

All space activities ultimately rely upon the ability to successfully track and communicate with spacecraft, whether crewed or automated. Tracking stations form the vital link between spacecraft and their ground controllers. When the United States decided that it would launch a satellite during the IGY, it also had to develop a global tracking station network in order to monitor the satellite's orbit and receive data from its onboard scientific instruments. Australia was crucially placed geographically to provide a site for tracking facilities that would contribute to maintaining this global contact network. Its geographic location, coupled with its excellent technical capabilities (acquired through missile tracking work at Woomera), suitable local environmental conditions and a government politically allied with the United States, combined to make Australia particularly attractive for the establishment of space tracking stations in support of the Vanguard project.

As NASA did not yet exist, the responsibility for developing the Vanguard satellite and its associated tracking network was assigned to the Naval Research Laboratory (NRL). The NRL developed a radio tracking system for Vanguard, known as Minitrack (for Minimum Weight Tracking). This system relied on a 'minimum weight' radio transmitter carried by the satellite. This beacon was tracked with radio interferometry, a technique based upon the use of two receiving points, and the comparison of the phases of the radio signals received separately by each, to provide an angle of direction to the satellite. Additional antennae, located with the tracking array, would receive telemetry from the satellite. The NRL realised that a tracking station located at Woomera would be ideal to check that the satellite had achieved orbit after launch, and the Australian Government was approached to allow the establishment of a Minitrack station there.

At the same time that the NRL was developing the Minitrack system, the Smithsonian Astrophysical Observatory (SAO) planned an optical satellite tracking network that would use the Baker-Nunn camera, a high-precision wide-aperture telescopic camera. By photographing moving satellites against the fixed-star background, the Baker-Nunn camera system enabled satellite positions to be determined with much greater accuracy than was possible using Minitrack. Because SAO Baker-Nunn observatories required cloud-free locations, the dry desert of South Australia provided exceptional observing conditions, making Woomera an excellent site for one of the planned twelve SAO stations around the world.

While the two tracking systems can be thought of as complementary — with Minitrack being the 'ears' and the Baker-Nunn camera the 'eyes' of satellite tracking — they required different local physical conditions for successful operation and Woomera was ultimately the only place in which the Minitrack and Baker-Nunn facilities were co-located.

The Australian Government agreed in 1957 to host both stations, and the agreements under which they were established set the precedents under which later NASA tracking stations in the country would be operated. The NRL and SAO would provide the equipment (and any necessary training for Australian staff) and pay for its transport to Australia. Australia would provide the land and cover the costs of the installation, maintenance and operation of the stations. An important requirement was that the stations be managed by Australian staff and operated on a day-to-day basis without direct interference from the US, although an American representative would be welcome to serve in an advisory capacity.

The Minitrack and SAO stations were housed in a facility established at Range G, a disused area located about 10km north-west of the Woomera township that had been used originally for short-range missile tests between 1949-53. The WRE also established a Satellite Centre (SatCen), located at WRE Headquarters in Salisbury, which acted as an information exchange, between the Smithsonian Astrophysical Observatory and the Moonwatch teams in Australia, poviding them with predictions of when satellites were due to pass over Australia so that they could be observed.

Although the Minitrack network became operational on 1 October 1957, just days before the launch of Sputnik 1, it was initially unable to track the satellite, which was broadcasting on the amateur radio frequencies of 20 and 40MHz, different from the frequency chosen for Vanguard. Yet, despite this issue, some rapid jury-rigging of suitable receivers enabled the Woomera Minitrack station to make its first track of Sputnik's signals about two days later. The SAO network, however, was not yet in operation, due to delays in the production of the Baker-Nunn cameras. SAO Woomera would not become operational until early March 1958, more than a month after the launch of the first US satellite, Explorer 1, and just in time for the long-delayed launch of Vanguard 1 on 17 March.

Between the launch of Explorer 1 and the end of 1960, when the Baker-Nunn and Minitrack networks would begin their transition into NASA's Satellite Tracking and Data Acquisition Network, the Australian space tracking teams, like their counterparts around the world, settled into a routine of observing the growing number of scientific and military satellites launched by the United States and the USSR. During this period, the first station chief at the Woomera Baker-Nunn observatory, who would soon be invited to work at the SAO headquarters in the US, developed techniques enabling the telescope camera to track a satellite for 8000km of its passage, representing almost 30 minutes of continuous photography. In August 1959, the Woomera observatory successfully employed this technique to photograph the satellite Explorer VI across a distance of 22500 kilometres. The first director at the Woomera Minitrack station would also be recruited to work in the United States, eventually becoming the Station Director at the Merritt Island Launch Area (MILA) tracking station at Kennedy Space Centre during the Apollo period. These men were among the first of many WRE officers who would find positions at NASA and in the US armed services over the next decade — a testimony to the training, skills and professionalism of the Weapons Research Establishment.

Operation Moonwatch–Citizen Science at the Dawn of the Space Age

Because it was predicted that the Vanguard satellite would appear very faint in the sky, just at the limit of naked eye visibility, optical tracking with the Baker-Nunn telescope camera faced the challenge of initially locating its high-altitude target. The SAO solution was to recruit a world-wide team of volunteer observers, particularly amateur astronomers, who would scan the skies at dawn and dusk (the periods when satellites

Above A view of the Minitrack tracking facility at its second location at Island Lagoon, near Woomera. (Photo courtesy of the Defence Science and Technology Group)

are at their most visible as they reflect the rays of the rising or setting sun), reporting any satellite sightings to the SAO. These sightings would provide approximate locations for the Baker-Nunn observatories to begin their precision tracking. This global volunteer program, known as Operation Moonwatch, was an example of what today would be termed 'citizen science'.

In Australia, Moonwatch was operated under the auspices of the Australian Academy of Science's National Committee for the International Geophysical Year, which appointed the Professor

Sputnik 1 and OTC

When Sputnik 1 was launched, the Overseas Telecommunications Commission (OTC), which was responsible for Australia's international radio and cable communications, had equipment available that could receive the amateur radio bands that Sputnik was using. The OTC international receiving station at Bringelly, near Sydney, detected the signals on 5 October, and soon after signals were detected at Rockbank, Victoria, the Coast Radio station in Hobart and Bassendean in Perth. By comparing the signal strengths and times of reception at each station, a good approximation of the satellite's orbital period was obtained. On 6 October, the first reported visual sighting of the satellite in Australia (possibly actually the carrier rocket stage, which had followed Sputnik into orbit and was much brighter than the satellite itself) came from Hobart.

OTC provided the Australian media with the first forecasts of when the satellite would be visible over different parts of Australia. These forecasts brought crowds out into the streets and parks, hoping to observe Sputnik pass over. Such was the Australian interest in the world's first satellite that OTC was besieged by calls as people reported satellite or UFO sightings. These included a call from a country newspaper reporter, who had heard that a 'flaming object' had plummeted to the ground near a town (though Sputnik would have been over Afghanistan at the reported time of the incident), and another from a person who declared that he could distinctly hear the Sputnik sounds emanating from his bed springs! One worried resident of Sydney supposedly took out the world's first insurance policy against being struck by a falling satellite, while rock-n-roll star Little Richard, touring Australia at the time, experienced a religious conversion as a result of seeing Sputnik.

ABOVE *When the USSR launched Sputnik 1 it took the world by surprise. Australia's Overseas Telecommunications Commission was among the first in the world to hear and track this very first satellite. Moonwatch volunteer groups in Sydney and Woomera were also early observers. (Illustration created by Glen Nagle)*

of Physics at the University of Adelaide as the National Moonwatch Co-ordinator. WRE's Satellite Centre also supported the Moonwatch activity by receiving and passing on observations to the US and disseminating copies of the SAO's Satellite Bulletin (which listed observations from tracking centres around the world), to assist local Moonwatch groups in locating satellites.

By October 1957, there were five Moonwatch groups preparing to observe the United States' first satellite: Sydney, Melbourne, Adelaide, Woomera and Perth. While most of these groups were associated with amateur astronomy societies, the Woomera Moonwatch was organised by staff of the WRE and worked in direct association with the Baker-Nunn Observatory. These five groups joined a global network of some 180 Moonwatch teams, eager to participate in one of the most exciting projects of the IGY.

Like the professional satellite trackers, Moonwatch groups were taken by surprise by Sputnik, since the launch of Vanguard was not expected until December at the earliest. However, as news of the satellite's launch broke, the Sydney and Woomera teams swung into action to observe on the night of 5 October, using the co-ordinates provided by OTC. Although cloudy skies prevented any sightings until 8 October, both Woomera and Sydney observed the satellite that evening, becoming the first Moonwatch groups in the world to do so. Sydney actually recorded three of the first four Sputnik visual observations: the Adelaide and Perth groups succeeded in observing Sputnik later in the week. Mount Stromlo Observatory, in Canberra, managed to photograph Sputnik on 8 October and both Sydney Observatory and Mount Stromlo photographed the satellite on 9 October.

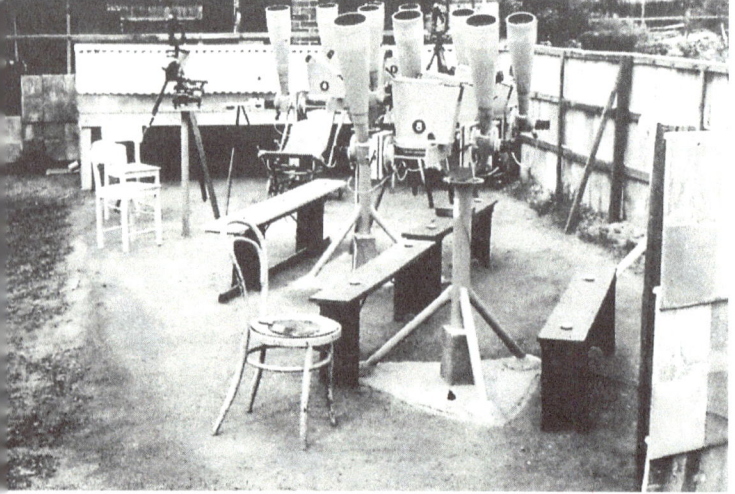

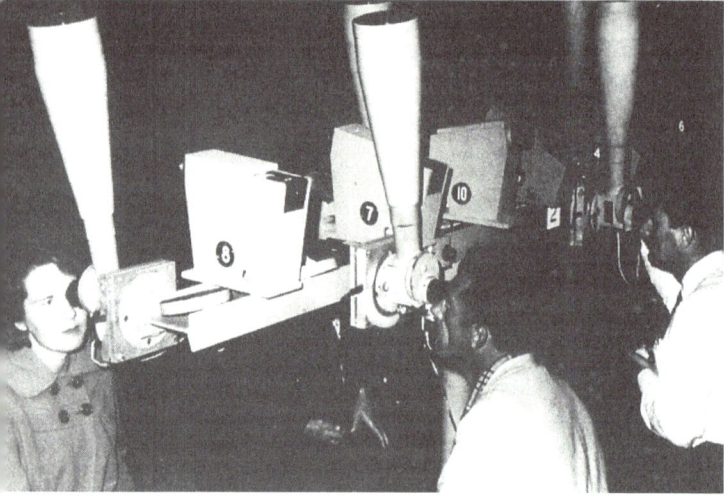

Above (top) A daytime view of the Perth Moonwatch observatory, showing the special telescopes designed for Moonwatch use and supplied by the Smithsonian Astrophysical Observatory. (Photo courtesy of the Astronomical Society of Western Australia)

Above (bottom) Members of the Perth Moonwatch team watching the skies to observe and track early satellites using the SAO telescopes. (Photo courtesy of the Astronomical Society of Western Australia)

By the end of 1957, Moonwatch stations had contributed approximately 700 observations of Sputnik 1 and 2 from groups in America, Australia, Chile, Japan and Curaçao. With the United States finally in orbit in early 1958, following the launch of Explorer 1 and then Vanguard 1, the Australian Moonwatch teams produced a steady flow of observations, providing valuable data to the SAO. After the entire Baker-Nunn network was completed in July 1958, Moonwatch groups were no longer critical for regular satellite tracking, but they continued to contribute to IGY research through participation in various projects, such as satellite re-entry watches and searching for satellites that had been 'lost' to the Minitrack and SAO stations.

By the end of 1958, all the Australian Moonwatch teams had received the designation of Prime A stations, based on the quality of their observational data. This rating enabled them to borrow more powerful telescopes and binoculars from the United States, to better participate in the SAO's special programs. One such project in which Australian teams successfully took part was the attempt to relocate the 'lost' satellite, Explorer IV. Launched in July 1958, the satellite's transmitters failed in early October, and Baker-Nunn stations lost sight of it in December of that year. An urgent request to Prime A stations to try to reacquire the satellite resulted in the Adelaide group finding it again on 1 January 1959, a fitting final achievement for the formal conclusion of Australian involvement in space tracking for the International Geophysical Year.

While the IGY had seen the establishment of space tracking projects in Australia, and allowed the WRE to acquire extensive experience in this vital space technology, the formal end of this special scientific period in December 1958 did not mean the cessation of space tracking operations. NASA, created in July 1958 to manage the United States' civil space program, took over the IGY tracking station networks, including the Moonwatch teams, in order to support its own space projects and an expanded role for Australia in civil space tracking would commence in 1961, as will be described in Chapter 4.

Sounding Rockets and the IGY

While the United States and the Soviet Union set themselves the goal of launching satellites during the IGY, other nations, with less ambitious goals and smaller budgets, made plans to launch sounding rockets to explore the upper reaches of the Earth's atmosphere, whose characteristics were still largely unknown at the time. Sounding (or geophysical) rockets do not put satellites into orbit, but carry packages of scientific instruments into the fringes of space, before falling back to the Earth.

The vast area of the Woomera Range was well suited to launching sounding rockets and recovering instrument payloads returned to the ground with parachutes. Consequently, both Britain and Australia made plans for programs of upper atmosphere studies as part of their contributions to the International Geophysical Year. They recognised that, for long-range missile testing, which would take rockets into the space environment, information on the basic upper atmospheric parameters of pressure, temperature, density, windspeed and wind direction was vital. Sounding rockets could, therefore, provide fundamental data that had both scientific and practical value.

The primary IGY sounding rocket program at Woomera used the British-developed Skylark rocket, which was capable of lofting payloads beyond 100km in altitude. The initial focus of Skylark research was to study the physics of the upper atmosphere, although this would greatly expand in the 1960s to also include infra-red, gamma ray and X-ray astronomy, as will be discussed in Chapter 3.

Australia commenced planning its IGY sounding rocket program in 1955, when the WRE initiated the High Altitude Research Project (HARP). The HARP vehicle was to be a rockoon, a combination of balloon and sounding rocket. HARP would use a large research balloon to carry a small rocket approximately 12km into the atmosphere, after which the rocket motor was fired by remote control to carry the payload to an altitude of 100km. The polythene HARP balloon was made in Australia to a British design, while the HARP sounding rocket was developed, under the guidance of the Propulsion and Aerodynamics Divisions of the WRE, by the Special Projects Division of the Fairey Aviation Company of Australasia Pty Ltd.

As Australian space science was in its infancy, six British universities provided payloads for the HARP project, using equipment similar to that planned for the first Skylark rockets experiments. The rocket was also equipped to collect and analyse air samples at high altitude. HARP test firings began in early 1957, but the balloon proved difficult to handle in Woomera's random wind gusts. It was eventually concluded that Woomera's uncertain winds made it unsuitable for rockoons and HARP was abandoned in 1958.

Australia's first successful home-grown sounding rocket, the Long Tom, was originally designed to test Woomera Range's instrumentation, before the

Above The Long Tom was the first successful Australian sounding rocket. Originally used to test Range instrumentation at Woomera, it was adapted for research use and made its first upper atmosphere research flight in 1959. (Photo courtesy of Defence Science and Technology Group)

commencement of the Black Knight trials program. After the failure of the HARP rockoon, Long Tom was re-purposed as a sounding rocket, becoming the first in a succession of Australian sounding rockets that would be developed in the 1960s and 70s specifically for research applications.

Long Tom was a two-stage vehicle based on surplus British rocket motors. It could accelerate faster and achieve a higher altitude (approximately 160km) than the Skylark. Because it carried a radar chaff dispenser for instrument tests, it could be used to complement a Skylark experiment that measured the speed and direction of upper atmospheric winds by ejecting and tracking various materials, such as radar chaff. The Australian National University also installed a micrometeorite detector on the rocket, to expand the range of data collected. Although flying for Range testing in 1958, the first Long Tom research flight was in May 1959.

In 1958, the WRE also developed a second small vehicle for use in testing the Woomera Range instrumentation, named Aeolus. Like the Long Tom, its two stages were assembled from surplus British rocket motors. Although not as powerful as the Long Tom, it could carry a payload up to 129km altitude. All the Aeolus flights included a chaff dispenser to measure wind patterns at altitudes from 30-77km. Flight number 5, on 23 May 1960 was particularly significant in that it tested a number of photo-theodolites, designed by the WRE to photograph the exploding chemical 'grenades' ejected from Skylark rockets during their atmospheric research.

Although Long Tom and Aeolus would soon be superseded by more advanced sounding rockets specifically designed for research use, they marked the beginning of Australia's significant upper atmosphere research program that would continue until 1975.

AUSTRALIA'S SPACE ROLE AT THE UNITED NATIONS

Australia's space-related activities during the IGY gained it an early seat at the table in United Nations Committee on the Peaceful Uses of Outer Space, where the country would assume a significant role for 35 years in the international oversight of world space activities.

Following the launch of the first Soviet and American satellites, there was growing concern in the international community that space might become yet another domain of Cold War competition between the East and West, or would be open to exploitation only by the limited number of countries that could muster the necessary resources to undertake space activities. In 1958, the United Nations established an ad-hoc Committee on the Peaceful Uses of Outer Space (COPUOS) to address these concerns. Australia was one of the 18 nations on this committee, represented by Dr David Forbes Martyn, a distinguished British-born Australian scientist, with a strong interest in upper atmosphere research. A leading figure in organising Australian participation in the International Geophysical year, in 1958 Martyn had established the Upper Atmosphere Section of CSIRO at Camden, New South Wales. Already widely known and respected internationally both as a scientist and a proponent of international co-operation in scientific research, Martyn became actively involved with COPUOS from the outset.

When the UN General Assembly established COPUOS as a permanent body in 1959, Australia was one of its 24 founding members, represented by Dr Martyn. The new permanent committee would have two subcommittees, the Scientific and Technical Subcommittee and the Legal Subcommittee, to deal with the complex issues that were anticipated to arise as space technology developed and space activities expanded. When these subcommittees met for the first time in 1962, Martyn was elected as Chair of the Scientific and Technical Subcommittee.

Why was an Australian selected to Chair this important subcommittee? In this tense period of the Cold War, Soviet Premier Kruschev was insistent that in each UN body, there should be a trio of committees or participants — one from the Eastern bloc, one from the West and a representative from a neutral country. This arrangement was followed with COPUOS, where the general subcommittee was chaired by an Austrian (a neutral), the legal subcommittee by a representative from Eastern Europe, while Australia was the Western nation chosen to chair the science and technology subcommittee, due to the level of space-related activities at Woomera.

Dr Martyn held the Chair until his death in 1970 and was succeeded by Professor John Carver, another prominent Australian space scientist, who held the post until his retirement in 1995. For any nation to hold the Chair of a UN committee for such a long period is extremely rare, and it is a testimony to both the scientific and diplomatic skills of Martyn and Carver that they were able to maintain this position over 35 years, even after Australia had ceased to be recognised as a significant spacefaring nation.

Australia's Chair position in UNCOPUOS enabled the country to exercise great influence with regard to UN decisions on space activities. Among the issues in which Australia took a leading role were the development of principles regarding the use of nuclear power sources in outer space and the promotion of education and training programs in remote sensing for developing nations. Australia also contributed to the formulation of the United Nations Treaty on Principles Governing the Activities of States in the Exploration and Use of Outer Space, including the Moon and Other Celestial Bodies (the 'Outer Space Treaty'), which forms the basis of international space law. This treaty is based in part on the 1959 international Antarctic Treaty, which Australia played a major role in framing.

Australia's active and significant participation in COPUOS, alongside its other space activities in the 1960s and 70s, contributed to its reputation as a leading spacefaring nation during the first two decades of the Space Age.

Woomera and the Aboriginal Community

The effect that the Anglo-Australian Joint Project might have on the desert-dwelling Indigenous Australians living in the Range area was of little interest to Britain's weapons test planners, anxious as they were to establish the Rocket Range. It was, however, extensively debated in the Australian parliament in the late 1940s. Although the planned long-range impact areas were located in the far north of Western Australia, the proposed centre-line of the range would overfly the Central Aboriginal Reserves and both the safety and culture of the tribal people living there were deemed to be at risk, both from crashing missiles and the intrusion of the observation posts, which were to be located in the Reserves.

Although the risk of injury was actually very low, due to the extremely low population density over the Range area (one person per 400 square kilometres on average) and the relatively low rate of rocket firings, both Aboriginal rights campaigners and the white landholders whose properties were encompassed by the Rocket Range protested against the danger from crashing missiles. As a result, the pastoral station communities were eventually provided with shelters into which they could retreat when test flights were being undertaken. Protection for the nomadic tribal Aboriginal groups of the Reserves was, however, not even considered. Fortunately, no missiles flew over the Reserves during the first decade of Woomera's operations: of the few that did overfly the Reserves during the ELDO program, only one impacted in the Reserves as the result of a malfunction.

At the time Woomera was being established, most Australians had little interest in Aboriginal issues, although a group of white Aboriginal rights campaigners did champion the cause of the Indigenous people in the early years of the Range. While some of their arguments were based on wildly incorrect surmises about the type of work that would be carried out at the Range, their protests were recognised and a government committee was convened in 1947 to examine the issues. Although this committee included state Aboriginal Affairs officers and a noted anthropologist, no representatives of the affected Aboriginal groups were invited to participate, as the attitude of the day was that white experts knew what was best for the Aborigines.

The findings of this committee, perhaps not unpredictably, were that the measures the government intended to take for the physical and cultural safety of the Aboriginal people in the Range area were adequate, particularly in view of the Range's supreme importance to the defence of Australia and the West. This quieted public fears and Woomera ceased to be an important Aboriginal issue, as far as white Australians were concerned, for the duration of the Joint Project.

Australian government policy until the last years of the Joint Project was for the gradual, but eventual, assimilation of Aborigines into white culture and Indigenous people living in the Woomera rangehead region at the time of its construction had already had considerable contact with white pastoralists and miners. However, tribal peoples with little previous contact with Europeans ranged across the Reserves over which Woomera's rockets would fly and culture shock, or cultural interference, during the construction and operation of the Range, was considered a greater threat to them than the rockets themselves.

At the time Woomera was established, there were approximately 1200 Indigenous people living in the Reserves and the adjacent Missions (run by

Right A downrange reconnaissance party encounters tribal Aborigines in 1950. The men are carrying woomeras to assist with their hunting. Despite initial concerns for the welfare of Indigenous peoples within the Range, they were deemed secondary to Woomera's importance for defence research. (Photo courtesy of Defence Science and Technology Group)

various religious groups) with the population rising to around 2000 during the peak years of Woomera's operation. Despite early concerns over observation posts being established across the Reserves, only one was ever constructed, the Giles Meteorological Station. There were, however, some incidents at Giles which vindicated the concerns of Aboriginal rights campaigners.

To safeguard the welfare of the Reserve Aboriginal groups, a corps of white Native Patrol Officers was established in 1947. These men displayed genuine concern for the Indigenous people, at times acting beyond their assigned responsibilities to effectively become welfare officers and social workers. The Native Patrol Officers assisted in policing the borders of the nuclear test areas as part of their duties and are known to have voiced their concern about the dangers of fallout to the population on the Reserves.

The establishment of Woomera restricted the local Kokatha people's access to their traditional lands, cutting them off from many of their cultural sites, which fell inside the prohibited areas. The western branch of the Kokatha also became separated from those living to the east of the Range, which led to a physical dispersal of the Kokatha people away from their traditional lands.

By the late 1950s, many Indigenous people had left the Reserve, drifting to townships or cattle stations. However, in the last years of the Joint Project, Aboriginal groups began to move back into the Reserve areas, with the growth of the Aboriginal land-rights movement. The diminished scale of activities at Woomera meant that the Range was not considered a hazard to these returning groups. The Kokatha and other Indigenous peoples with traditional connections to the land encompassed by the Woomera Prohibited Area began to press for access to their traditional lands under Native Title legislation in the early 1990s. Their campaign has gradually met with success, regaining a level of access to the Range, enabling them to practise traditional hunting and gathering activities and cultural activities at some significant sites.

In 2014, the Native Title holders were granted the pastoral leases of three properties that had originally fallen inside the Woomera Range: Andamooka, Purple Downs and Roxby Downs Stations, which now form the heart of the Kokotha Pastoral Company. This has allowed members of the Kokatha community to once again live and work on part of their traditional lands. They continue to campaign to have areas damaged by mining and other activities on the Range rehabilitated to their natural state, and seek further access to the Range to prevent damage to culturally significant areas.

BACKGROUND IMAGE THIS PAGE *The versatile Skylark rocket design used many different motor combinations. Here a Skylark with a Cuckoo bost motor soars skyward. (Photo courtesy of Defence Science and Technology Group)*

CHAPTER 3

Liftoff: Going Up from Down Under

The International Geophysical Year, by initiating space-related projects at Woomera, set the stage for Australia to become a player in the space arena in the 1960s. As the Space Race developed between the United States and Soviet Union, the international prestige it garnered for both nations and a growing awareness of the benefits of satellites for military and civilian use encouraged other nations, especially Great Britain and those of Western Europe, to embark upon their own space programs. With its remote desert location, Woomera was ideal for testing satellite launchers as well as sounding rockets and by the mid-1960s it had joined the ranks of the world's 'spaceports'. Locally developed sounding rocket projects would pave the way for Australia to launch its own satellite into orbit in 1967.

OPPOSITE ABOVE *A Blue Streak rocket lifts off on a test flight for the first stage of the ELDO Europa launcher. (Photo courtesy of Defence Science and Technology Group)*

OPPOSITE BELOW *Europa F8 lifts off on an unsuccessful attempt to place a satellite in orbit in July 1969. (Photo courtesy of Defence Science and Technology Group)*

BACKGROUND IMAGE THIS PAGE *An early view of the sounding rocket launch area at Woomera. The tall Skylark launch tower was made from war surplus Bailey bridge segments. (Photo courtesy of Defence Science and Technology Group)*

BLUE STREAK: FROM MISSILE TO SPACE LAUNCHER

In 1953, the British government reversed its policy on long-range missiles, having decided that, with the Cold War deepening, including such weapons in its nuclear arsenal was now imperative. This decision led to the Blue Streak program, initiated in 1955, intended to develop a long-range ballistic missile capable of hitting a target in Eastern Europe or the USSR if launched from either Britain or the Middle East (where Britain still had colonial territory).

Guided and propelled only during the initial phase of its flight, the trajectory of a ballistic missile is largely determined by gravity and atmospheric drag, like that of a bullet or artillery shell. The flight trajectory of a ballistic missile with a range of a few thousand kilometres carries it into the space environment before it re-enters the atmosphere over its target territory. Almost all the world's early civilian space launch vehicles were derived from modified intercontinental ballistic missiles (ICBMs).

In April 1956, the Australian government formally accepted the British proposal for the Blue Streak missile, and a smaller associated research rocket known as Black Knight (further discussed in Chapter 5), to be tested at Woomera. The proposed time scale of the project was for the first Black Knight launch in 1958 and the first Blue Streak test in 1960. Thus, ten years after the initial reconnaissance to establish a location for Woomera, a program to develop the Rocket Range to the extent originally envisaged in 1946 was finally set in train – a program that would, within another decade, expand Woomera's role from a weapons testing range to a spaceport.

The development and testing of Black Knight and Blue Streak would require a vast expansion of the Range. The facilities at Woomera had become

BELOW Launch Area 6 at Lake Hart was originally developed for the Blue Streak missile program before being used for the ELDO launcher project. In the foreground, a Europa vehicle awaits launch on pad 6A. Launch site 6B, never made operational, can be seen in the background. (Photo courtesy of Defence Science and Technology Group)

concentrated in a relatively small area, but now a much larger part of the prohibited area would have to be put into use. This was a major engineering assignment requiring huge constructions at the planned Blue Streak launch sites at Lake Hart and the installation of tracking, measuring and recording instruments across central Australia to the Talgarno impact area in Western Australia. An enormous upgrading program was initiated: new roads were laid over hundreds of kilometres of untouched country and water pipelines, power supplies and telephone lines were extended to the new launch sites.

However, the Blue Streak missile became obsolete as a weapon while it was still under development and the UK cancelled the program in April 1960, ironically while the first missile was en-route from Britain to Woomera for tests. The sudden decision was taken without any real consultation with Australia. After so much money had been expended on the missile's development, and Australia had invested considerable resources on new facilities at Woomera, the cancellation caused a major outcry in both countries and severely embarrassed the British and Australian governments.

Anxious to salvage the expenditure that had been made on the Blue Streak, Britain investigated the possibility of turning the missile into the first stage of a satellite launch vehicle. A satellite launcher based on the Blue Streak and Black Knight had been suggested as early as 1957 and the idea was now put forward as the basis for a Commonwealth satellite launcher, to be developed and used by Britain and other Commonwealth nations. However, no Commonwealth country, including Australia, expressed any interest in this project.

The European Launcher Development Organisation and the Europa Launcher

When the Commonwealth satellite launcher proposal failed to attract any interest, Britain sought to recover from the cancellation debacle by offering the Blue Streak as the first stage of a satellite launch vehicle to be developed by a consortium of European nations. France expressed interest in this project and by 1962, Belgium, the Federal Republic of Germany (West Germany),

RIGHT The official logo of the European Launcher Development Organisation. The acronym CECLES used in the logo stands for the French name of the organisation: Conseil européen pour la construction de lanceurs d'engins spatiaux. (Courtesy of European Space Agency)

Italy and the Netherlands had also agreed to participate, leading to the formation of the European Launcher Development Organisation (ELDO). Although ELDO was set up between 1961 and 1962, it did not formally come into being until 29 February 1964, because of the complexity of the international negotiations needed to ratify its charter.

ELDO had as its objective the development of an independent, non-military European satellite launch vehicle, which was to be called Europa. With the Blue Streak as a first stage, the launch facilities already in place in Australia were the obvious site for ELDO launch operations and the nation consequently took another step forward in space activity, becoming the only non-European member of ELDO. At the insistence of the Commonwealth Government, Australia was considered a full, but non-paying, member of ELDO, contributing the Woomera facilities and their operation in lieu of financial commitment.

The proposed new launcher to be used in the first phase of the ELDO program was dubbed Europa 1. The vehicle consisted of three stages, each provided by a consortium member: the first stage was a modified Blue Streak provided by Britain; France provided the second stage, called Coralie (named in part because Coralie rhymed with Australie, the French word for Australia); and West Germany the third stage, known as Astris. The test satellite that the Europa vehicle was to launch was developed under the leadership of Italy. It would carry instruments for tracking, data recording and telemetry, as well as a radio control system. The Netherlands and Belgium were responsible for the development of telemetry and guidance systems.

Although Australia was not directly involved in the design and technical development of the Europa, the launch, tracking and monitoring

ABOVE A British Blue Streak rocket awaits testing in the UK, before being shipped to Australia. The distinctive zigzag pattern on the body allowed accurate measurement of how the rocket rolled after launch. (Author's collection)

facilities were managed for ELDO by the WRE and range safety responsibility rested with a WRE officer. A member of the WRE computing team also made a unique contribution to the rocket's testing: the application of the distinctive zigzag pattern on the Blue Streak stage. This enabled very accurate measurements to be made as the rocket rolled after leaving the launchpad. Using the pattern, the cameras that tracked the launch could easily measure if, and how far, the rocket had rolled, depending on where that diagonal was relative to the top and bottom stripes. The WRE also designed and oversaw the construction of a downrange tracking station at Gove (Nhulunbuy) in the Northern Territory, which was operated for ELDO by Hawker de Havilland Australia.

Two launch sites had been constructed for Blue Streak missile tests at Woomera's Launch Area 6 (LA6) on the edge of the Lake Hart salt lake. Of the two, only 6A was actually completed and used for the Europa 1 launches. Between 1964 and 1970, ten launches in the Europa 1 development program were made from Woomera, with five regarded as successful. Although the Blue Streak first stage experienced few technical problems in flight, there were repeated failures of the second and third stages: despite three attempts to launch the Italian test satellite (named STV or Satellite Test Vehicle), the Europa rocket never successfully placed a satellite into orbit.

Table 3.1 Europa launches in Australia

Flight	Launch date	Vehicle configuration	Remarks
F1	5 June 1964	1st stage only	Partial success
F2	20 October 1964	1st stage only	Successful
F3	22 March 1965	1st stage only	Successful
F4	24 May 1966	Active 1st stage with dummy upper stages and dummy satellite	Flight terminated at 136 seconds after launch
F5	13 November 1966	Active 1st stage with dummy upper stages and dummy satellite	Successful
F6/1	4 August 1967	Active 1st and 2nd stages, dummy 3rd stage and satellite	2nd stage failed to ignite
F6/2	5 December 1967	Active 1st and 2nd stages, dummy 3rd stage and satellite	1st/2nd stage separation failed
F7	30 November 1968	All stages active, live satellite	3rd stage exploded
F8	3 July 1969	All stages active, live satellite	3rd stage exploded
F9	12 June 1970	All stages active, live satellite	All stages successful, Satellite failed to orbit

Note: the F4 flight was terminated because it was thought the rocket had gone off course. In fact, the rocket was functioning perfectly and it was an error in the flight predictor which incorrectly indicated a flight path deviation.

CHAPTER 3 – Liftoff: Going Up Space from Down Under

Above A perfect Europa rocket lift off, but the attempt to place a satellite in orbit was unsuccessful. (Photo courtesy of Defence Science and Technology Group)

Right Diagram of the main components of the ELDO Europa 1 satellite launch vehicle.

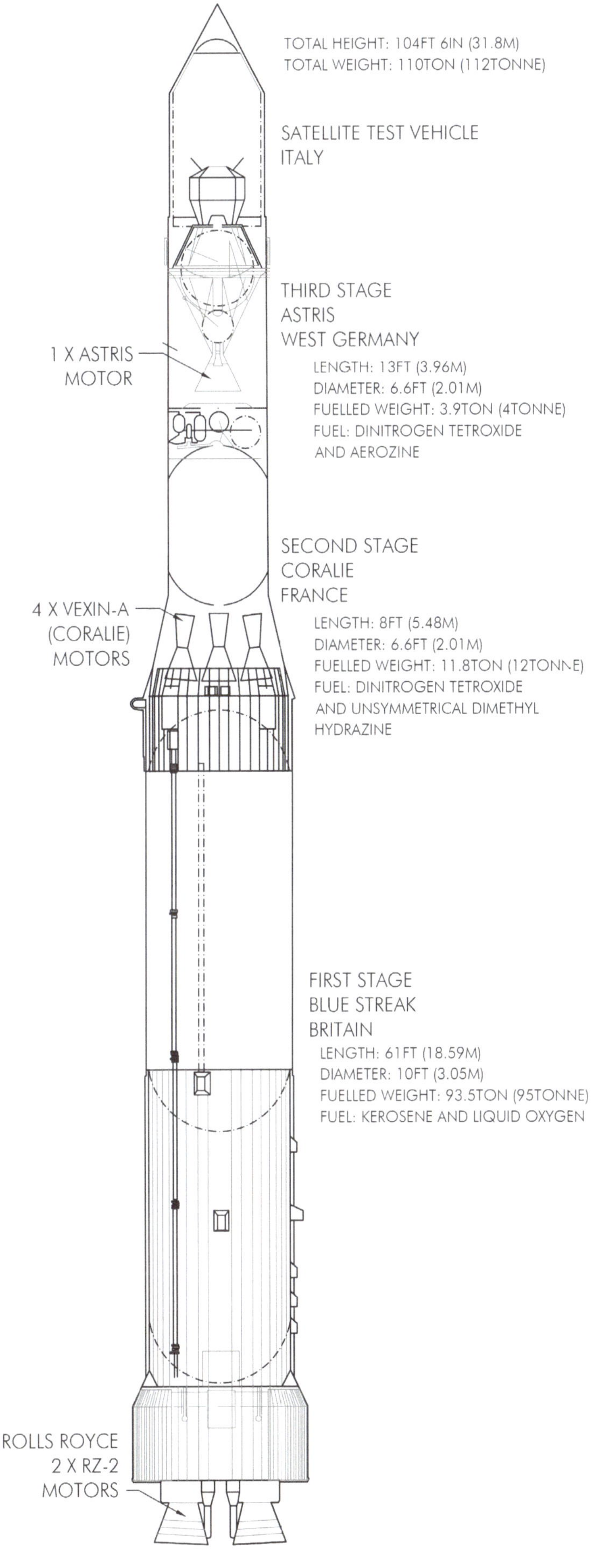

TOTAL HEIGHT: 104FT 6IN (31.8M)
TOTAL WEIGHT: 110TON (112TONNE)

SATELLITE TEST VEHICLE
ITALY

THIRD STAGE
ASTRIS
WEST GERMANY
LENGTH: 13FT (3.96M)
DIAMETER: 6.6FT (2.01M)
FUELLED WEIGHT: 3.9TON (4TONNE)
FUEL: DINITROGEN TETROXIDE AND AEROZINE

1 X ASTRIS MOTOR

SECOND STAGE
CORALIE
FRANCE
LENGTH: 8FT (5.48M)
DIAMETER: 6.6FT (2.01M)
FUELLED WEIGHT: 11.8TON (12TONNE)
FUEL: DINITROGEN TETROXIDE AND UNSYMMETRICAL DIMETHYL HYDRAZINE

4 X VEXIN-A (CORALIE) MOTORS

FIRST STAGE
BLUE STREAK
BRITAIN
LENGTH: 61FT (18.59M)
DIAMETER: 10FT (3.05M)
FUELLED WEIGHT: 93.5TON (95TONNE)
FUEL: KEROSENE AND LIQUID OXYGEN

ROLLS ROYCE 2 X RZ-2 MOTORS

As a multi-national program, with different, often competing political and economic priorities among its members, ELDO experienced administrative and economic difficulties as the program progressed and several technical problems arose. The launcher underwent many modifications and costs began to escalate. Development and launch delays pushed the program schedule back and enthusiasm for the Europa vehicle began to wane when ELDO's sister organisation, the European Space Research Organization (ESRO), conducted successful satellite programs launching with the United States.

As early as 1965, ELDO began to rethink its launch site requirements. One of the consortium's intended purposes was to launch telecommunications satellites for its members, in addition to scientific satellites. Although scientific satellites are often placed in polar or inclined orbits, which could be readily attained from Woomera, communications satellites are best placed in much higher geostationary orbit, where they can broadcast across a wide swath of the Earth's surface. To place a satellite into geostationary orbit efficiently, a launch site near the equator is best, because the Earth's own rotation will add significant velocity to a rocket launched there toward the east.

However, at approximately 31°S, Woomera was too far south of the equator for a launch there to gain any useful velocity by firing to the east. In addition, an eastward launch was potentially dangerous to the more densely populated coastal regions of Australia. Although it would have been possible to achieve equatorial orbit from Woomera, 20 per cent more energy would have been required than for a launch from a site near the equator, making such a launch uneconomical.

Also in 1965, Britain had begun to reconsider its role in ELDO, for economic and political reasons. In 1966 it reduced its original funding commitment of 40 percent of costs to 27 percent. In 1968 Britain decided to withdraw from ELDO and instead pursue its own satellite launcher, the Black Arrow. With France now the consortium's largest funder, and with a growing focus on communications satellites, the remaining ELDO members decided to pursue further launches from France's equatorial launch site at Kourou in French Guiana, now the site of the European Space Agency's launch facilities. The last ELDO launch at Woomera occurred on 12 June 1970, without the program ever having successfully put a satellite into orbit.

Despite being the largest space project conducted at Woomera, today little evidence of the ELDO project remains there. The Lake Hart facilities were scrapped in the 1970s, leaving only the massive concrete launch pad bases, chipped and scarred from later being used for target practice during Defence training exercises. The Gove tracking station, too, was demolished, with the antennas transferred to the WRE laboratories at Salisbury for research work.

Early Proposals for Equatorial Launch Facilities in Australia

Woomera's less-than-ideal location for launching geostationary satellites was recognised as early as 1960, when the possibility of an equatorial launch facility on Cape York was raised during Britain's initial attempts to interest Australia in the development of a satellite launcher. In 1962, France briefly looked to Australia as a possible location for its new national launch facility: Broome and Darwin were both considered early candidate sites before France decided to base its new launch centre at Kourou in its South American territory of French Guiana.

In 1964, expatriate Australian physicist Dr Philip K. Chapman (who would go on to become the first Australian-born person to be selected as an astronaut, in 1967) put forward the suggestion of establishing an international equatorial launch facility on Manus Island, off Papua New Guinea. Chapman's idea was that Australia should build and manage an equatorial spaceport that would be open to any country as a launch site, even the Soviet Union. This somewhat radical suggestion for the time did not, however, find favour with the Australian Government and the proposal was never adopted.

With its own plans for future geostationary launches requiring an equatorial launchsite, ELDO commissioned, in early 1965, five studies by different member nations (UK, France, the Netherlands, Germany and Australia), to investigate possible options. The Australian Government was anxious to retain ELDO launch facilities within the country, concerned that, should ELDO move offshore, and with Britain's waning interest in either satellite launchers or major weapons development, the nation could be left with Woomera as a 'white elephant' — a facility for which the country had no use of its own (since it was not considering the establishment of an Australian space program), and into which it had poured millions of pounds in development funds. The establishment of an ELDO facility at Darwin, or elsewhere in northern Australia, would also contribute to the government's ongoing priority for development of the country's northern regions.

The WRE proposed a modest, but potentially expandable, base that would be located approximately 35km south-east of Darwin and operate as an adjunct of the Woomera Range. An initial single launch emplacement was suggested, with options for later expansion as required. However, with Britain's influence in ELDO diminishing, and France assuming the role of major contributor to ELDO's funds, the selection of its Kourou launch facility in French Guiana as ELDO's future equatorial launch base is perhaps not surprising. Australia's loss was indeed France's gain, with Kourou ultimately becoming the spaceport of the European Space Agency, which developed after the demise of ELDO in the early 1970s to become one of the world's major space agencies today.

Black Arrow: Britain's National Satellite Launcher

One reason for Britain's withdrawal from ELDO was its decision to make a new attempt at developing an independent national satellite launcher through the Black Arrow project. It was to be the last major rocket program undertaken at Woomera, up to the present day. Developed from the reliable Black Knight, the Black Arrow was a three-stage rocket, the first two stages fuelled by kerosene and high test peroxide (a concentrated form of hydrogen peroxide), with a Waxwing solid rocket as the third stage. Although the first launch was a failure, three further Black Arrow flights were successfully conducted at Woomera. The rocket's final flight, on 28 October 1971, lofted into orbit Prospero, the second of only two satellites so far (the first was Australia's WRESAT) to be orbited successfully from the Range.

Despite the successful launch of Prospero, the Black Arrow program was closed down, as Britain decided that it was more economical to purchase launches on American Scout rockets than to maintain its own launcher program. This decision actually made Britain the only nation in the world to have achieved an independent national satellite launch capability and then discontinued it.

Above The final flight of the Black Arrow rocket placed the satellite Prospero into orbit. It was named for the Shakespearean wizard in The Tempest who gave up magic – an allusion to the fact that the Black Arrow program was cancelled shortly before the satellite was launched. (Photo courtesy of Defence Science and Technology Group)

Exploring the Fringes of Space — Sounding Rocket Programs

As described in Chapter 2, sounding rocket firings commenced at Woomera during the International Geophysical Year and these important research rockets would continue to be flown from the sounding rocket facility at Range E until the dying days of the Joint Project. While not receiving the same level of public attention as the satellite launcher projects, these comparatively cheap scientific research vehicles provided regular access to the near-space environment for universities and government research agencies, and for small research projects of the major space agencies. The low cost of sounding rockets compared to satellites meant that the same types of instruments could be flown repeatedly, enabling large data sets to be compiled over time.

The British and Australian sounding rockets programs at Woomera provided valuable information on the physics of the upper atmosphere (such as the level of ozone, the structure and composition of the ionosphere and the measurement of high-level winds) and contributed to new discoveries in infra-red, gamma, ultraviolet and X-ray astronomy, as well as early investigations into remote sensing techniques.

BELOW The most powerful version of the Skylark rocket launched in Australia used a Goldfinch boost motor to reach an altitude of 270km. This type of Skylark was primarily used for space astronomy flights. (Photo courtesy of Defence Science and Technology Group)

Skylark

Following the first flight on 13 February 1957, some 250 British Skylark rockets would be launched at Woomera over the next 22 years, serving not only UK researchers, but also Australian, American and European space scientists: NASA made its first Australian sounding rocket campaign in 1961, with four Skylark flights at Woomera. A versatile design that could be upgraded through the use of different rocket motor combinations, the most advanced version of the Skylark used in Australia could carry around 200 kilograms of instruments up to an altitude of 270 kilometres. Although Skylark launches at Woomera ceased in 1979, as the Joint Project came to an end, the rocket was so successful that it stayed in operation at European launch sites until 2005.

With its ability to point at the Sun, Moon or a star, the Skylark made an excellent platform for space astronomy. Arguably the most important projects undertaken with Skylark rockets at Woomera were the X-ray and ultraviolet astronomy programs conducted by University College London and Leicester University and by Australian researchers from the Universities of Adelaide and Tasmania. The X-ray astronomy program was of particular significance: the first good-quality X-ray images of the solar corona and the first X-ray surveys of the Southern Hemisphere sky were captured with instruments flown on Skylarks. Major achievements of the X-ray astronomy research included the discovery by the Australian team, in 1967, of a powerful new X-ray source near the Southern Cross and their demonstration that X-ray stars varied in strength from day to day. Another major achievement was the British team's accurate determination, in 1972, of the location of deep space X-ray source GX3+1. Australian involvement with research using Skylark continued until the last years of the program, with the University of Adelaide providing instrument packages for Skylark ultraviolet and X-ray studies in 1977.

Australian Sounding Rocket Programs

Alongside the Skylark program, Australian researchers from the WRE and the University of Adelaide carried out their own upper atmosphere research program, which drew on the WRE's experience in propulsion systems and the development of robust instrumentation and camera systems. Following on from the Long Tom and Aeolus vehicles, in 1960 the WRE Project Studies Group made a conscious decision to design sounding rockets tailored for Australian research projects. These programs were intended to complement the research already being undertaken by British and Australian researchers using Skylark rockets, thus providing a broader range of data that was useful for geophysical and meteorological research as well as the Range's specific operational needs.

While the first WRE sounding rockets were constructed using surplus British rocket motors, the aim was to increase their Australian-made content over time and ultimately produce fully Australian-made sounding rockets. More than 400 Australian-designed and built sounding rockets, using various combinations of British and Australian-made rocket motors, would be launched up to 1975, allowing the accumulation of significant long-term data sets relating to near-space environmental conditions over Australia. This data was significant in improving the understanding of Australia's meteorological conditions.

The first purpose-designed Australian sounding rockets were the High Altitude Density (HAD) and High Altitude Temperature (HAT), developed by the WRE to explore the atmosphere in regions around 75-120 kilometres. Both were made from combinations of British rocket motors. HAT rockets used dropsondes (instrument packages dropped from the rocket, which fell to Earth on a parachute) to measure atmospheric pressure, temperature and ozone content as they descended. To further other Australian research already being carried out with Skylark, an ozone measuring device was added to the dropsonde in 1963, which measured the absorption of ultra-violet radiation as a means of determining ozone levels. The data from this research also fed into work being carried out in the HAD program, to investigate molecular oxygen in the upper atmosphere. 64 HAT vehicles were flown between 1960 and 1970, when the rocket was phased out in favour of the 'all-Australian' Kookaburra.

The High Altitude Density rocket (HAD) was developed to participate in research on the density of the upper atmosphere. The glowing clouds of chemicals released by exploding Skylark 'grenades' not only provided information on the temperature of the upper atmosphere, they could also be tracked and photographed to provide information on atmospheric density at different altitudes. The WRE developed photometers and camera systems to record the behaviour of these clouds, and in 1962 decided to commence a major project of glow cloud observation, so that windspeeds and atmospheric density could be deduced from their expansion.

WRE also decided to complement the Skylark work by developing its own rocket and density payload that would concentrate on the lower altitudes between 40 and 90km. This would become the basis for the WRE's longest continuous program of upper atmospheric research, commencing with HAD launches in 1962. Between 1961 and 1969, 118 HAD rockets were launched, before it was phased out in favour of the Kookaburra and Cockatoo vehicles.

One of the main HAD rocket payloads was the series of 'falling sphere' experiments, later continued with Kookaburra rockets. Between 1962 and 1975, these payloads were used to provide measurements of air density, atmospheric temperature, and wind directions and velocities. These experiments used a two metre balloon made of aluminised polyester film, which was packed with 'French chalk'. After the balloon was released from the rocket, it fell through the atmosphere until it collapsed at an altitude of around 30 kilometres. As it fell, the balloon was tracked by radar and the 'French chalk' was released from it, appearing as an expanding cloud with the bright sphere at the centre.

In 1964, a series of 'falling sphere' HADs were launched simultaneously at Woomera and Carnarvon, in Western Australia, in order to investigate variations in atmospheric density and winds over the western half of Australia. It was the first time in Australia that rockets had ever been launched outside the Woomera Range and the firings were assisted by the NASA tracking station at Carnarvon. By 1966, the HAD experiments had become monthly, providing routine measurements that formed a significant consistent set of seasonal atmospheric data. These monthly measurements continued until 1975, when the program ceased. One important experiment in this series, in 1965, launched ten rockets over a period of 24 hours, in an ambitious program to examine the effects of atmospheric changes between day and night.

HAD rockets were also used in a collaboration between WRE and the University of Adelaide, to investigate the chemical composition of the upper atmosphere, particularly the profiles of molecular oxygen and ozone. Already working on Skylark projects, from 1962 Adelaide University Professor of Physics, John Carver, collaborated with the WRE on the development of UV and X-ray ionisation detectors that could be flown on HAD vehicles to measure different atmospheric constituents. Within a few years, Carver's upper atmosphere work with HAD would pave the way for Australia's first satellite, WRESAT, as will be discussed below.

The effort needed to develop WRESAT within its very tight schedule was so intense that, except for the monthly HAD launches, all other Australian upper atmosphere research at Woomera virtually ceased during 1967. This break allowed the WRE to re-assess its sounding rocket programs and consider their future development once regular firing resumed in 1968. HAD and HAT were phased out in favour of a new generation of sounding rockets named after Australian birds, using Australian-made rocket motors that were the technological descendants of the Malkara missile's Marloo motor.

The first, and the largest, sounding rocket to use an Australian-made motor was Aero High. First launched in 1964, it was designed for another WRE program intended to complement research being undertaken with Skylark. Researchers were interested in the way in which the atmosphere reacted with different chemicals, especially when they produced chemiluminescent (glowing) reactions, leading to the Aero High being developed for the dual purpose of studying chemiluminescent reactions and measuring various parameters of the upper atmosphere between 65 and 200km.

The Aero High's second stage was an Australian-developed motor called Vela. It was made from a modified British motor casing, filled with an Australian-developed propellant. Although the first flight of the Vela motor was only partially successful, the Aero High vehicle would return after WRESAT, making 12 flights between 1968-1972, some conducted in conjunction with Kookaburra and Cockatoo experiments. These flights carried grenades which ejected lithium vapour and other chemicals into the atmosphere to create glowing clouds that enabled the detection of upper atmosphere winds and turbulence.

LEFT *'Exploded' view of a HAD sounding rocket, showing its major components, including the parachute used to return scientific instruments safely to the ground after a flight. (Photo courtesy of Defence Science and Technology Group)*

Aero High's Vela motor paved the way for local production of rocket motors to begin in earnest once the WRESAT project had been completed, forming the basis of an entire new family of Australian sounding rockets. Although some British motors continued to be used, there were ongoing efforts to replace them with Australian-made equivalents into the last days of the WRE sounding rocket program.

The new Australian rockets were named after native birds: Kookaburra, Cockatoo, Lorikeet and Corella. They were simple, functional vehicles tailored toward the requirements of particular research programs. Their new motors, named after constellations of the southern sky, were filled with WRE-developed solid propellants. The Musca motor was entirely WRE-developed, while the others, like the Vela, used British motor casings filled with improved Australian propellants. These motors were combined in various ways to produce the new sounding rockets: different propellants were used in different versions of the motors, creating different Marks of some rockets, providing different performance suitable to particular research projects.

HAT rockets were superseded in 1970 by Kookaburra launchers, which were also used for dropsonde research up to an altitude of 75 kilometres. In 1973, the interim Lorikeet vehicle was introduced, followed by the more powerful Kookaburra Mk 2. More than 100 Kookaburra and 15 Lorikeet rockets were fired before the shutdown of Australian sounding rocket research. HAD vehicles were replaced by the Cockatoo, a more aerodynamically efficient sounding rocket, which could reach altitudes of up to 130 kilometres. Like Aero High, Cockatoo rockets were used to carry out lithium-trail experiments, in which clouds of glowing lithium vapour were released into the sky to provide optical measurement of winds and turbulence in the atmosphere at around 80 kilometres. Other Cockatoo research programs included measurements of ozone concentration, ultraviolet radiation and other conditions in the ionosphere. More than 60 Cockatoo firings were made between 1970 and 1975. Shortly before the conclusion of this research at WRE, the Aero High was superseded by the Corella, which saw only two flights before the end of the program.

Above Aero High was the first Australian sounding rocket to use an Australian-developed motor. The Vela rocket motor powered its second stage. (Photo courtesy of Defence Science and Technology Group)

Between 1970 and 1972, the WRE also conducted a short program of meteorology-related atmospheric research using an American sounding rocket called HASP (High Altitude Sounding Projectile), which released clouds of metallised strips at an altitude of 60 kilometres. These strips could then be tracked by radar, indicating wind-speeds and directions. Almost 100 HASP rockets were launched during this program.

In 1970, after a decade of sounding rocket research at the WRE, an Interdepartmental Committee investigating activities at Woomera in the light of the expected wind down of the Anglo-Australian Joint Project reviewed the upper atmospheric research at Woomera. It directed that future research should be slanted more toward ionospheric observations with possible defence relevance. Following the Committee's report, there was a strongly held view that research which did not contribute directly to Australia's defence capabilities, should not be supported by Defence funds. Sounding rocket activity without obvious defence applications could not be justified in a tougher economic climate and the launch programs ceased in 1975, with the Upper Atmosphere Research Group being formally disbanded in 1976.

During the period 1960-1976, the Australian sounding rocket research conducted at Woomera contributed significantly to knowledge of the upper atmosphere and it is unfortunate that the research was terminated just at the point where it was reaching maturity and starting to yield long-term results, both scientifically and in the level of technical expertise in all aspects of sounding rocket construction that had been developed within the different branches of the WRE and Department of Supply.

Following the termination of the Skylark research in 1979, only spasmodic sounding rocket launches were conducted at Woomera for the next 20 years, such as in 1987, when NASA and West German teams utilised the Range to launch sounding rockets to study Supernova 1987A (detected that year in the Large Magellanic Cloud). The Australian Space Research Institute also carried out two firings of its AUSROC II sounding rocket development project in the early 1990s. However, since the beginning of the Twenty-first Century, the sounding rocket facilities have taken on a new lease of life supporting defence and civilian supersonic and hypersonic aerospace research projects. The HyShot and HIFiRE hypersonic research programs (see Chapter 5), together with the Japanese NEXST supersonic aircraft development tests, have used sounding rockets to launch experimental scramjet and supersonic aircraft models. The requirements of these projects have led to a significant refurbishment of the sounding rocket facilities and they remain operational and available for future scientific sounding rocket and hypersonic research use.

BELOW A Cockatoo rocket, mounted horizontally on its launch rail, before the launcher is elevated into firing position. (Photo courtesy of Defence Science and Technology Group)

WRESAT: Australia's First Satellite

In late 1966, Australian upper atmosphere research took the step from sounding rocket to satellite as the result of a fortuitous combination of circumstances, which allowed Australia to launch its first satellite in November 1967, a not insignificant achievement just ten years after the beginning of the Space Age.

In 1966, Britain and Australia were involved in a US-led project to investigate the physics of high velocity warhead re-entry into the Earth's atmosphere, dubbed Project SPARTA (Special Anti-missile Research Tests, Australia), which will be further discussed in Chapter 5. This program used American Redstone boosters (the same type of rocket that had been used to launch the United States' first astronaut, Alan Shepard) with two small upper stages, to launch its re-entry test heads. Ten Redstone rockets were brought to Woomera for this program, but by the latter part of 1966 it was obvious that only nine would be needed to complete the research.

Three senior WRE officers realised that this spare vehicle could make an ideal satellite launcher (and, in fact, a modified version of the Redstone had been used to launch the first American satellite, Explorer 1). They became excited by the possibility of extending the WRE's upper atmosphere research into orbit with the development of an Australian satellite. An informal approach to the US SPARTA team received a positive response: instead of being shipped back to America, the vehicle would be formally offered to Australia. In addition, the US team offered to prepare and fire the Redstone for the satellite launch. However, taking advantage of the US offer placed the project on a very tight schedule, because it meant that the satellite would have to be ready for launch by the end of 1967, when the SPARTA project would be complete and the Americans returning home. Thus, in the incredibly short span of only 11 months, Australia's first satellite, WRESAT (WRE Satellite) was designed, constructed, tested and finally launched on 29 November 1967.

To proceed with the project, Australian government approval would be necessary. However, the Liberal government of the period had not shown any particular support for developing an Australian space program and had already declined to take up an earlier US offer to launch an Australian-built satellite. Fortunately, the Minister of Supply obtained Cabinet approval for the satellite project to proceed, primarily on the basis that WRESAT would offer Australia the chance to gain international prestige and become a member of the 'Space Club', at a very low cost: the launch vehicle and launch services were being provided free; the University of Adelaide was contributing part of the cost of the satellite's experiment package; NASA and ELDO agreed to provide free tracking of the satellite, and many other expenses could be absorbed within the Joint Project budget.

With such a short development period available for the satellite, WRESAT became an excellent example of Australian skill at 'making do' in order to have the satellite ready in time. With the WRE's existing expertise in upper atmosphere research, it was decided that WRESAT's scientific payload should consist of instruments very similar to those already used in the sounding rocket programs, as these could be easily and quickly developed. It also meant that the research already undertaken with HAD and HAT would be extended, by providing comparison data from orbit.

The University of Adelaide team under John Carver, together with the upper atmosphere research group at the WRE, developed a suite of instruments to detect and record solar radiation at three of the wavelengths that most directly influence the temperature and composition of the upper atmosphere. There were two sets of three ion counters to measure ultra-violet radiation, X-ray counters and a photocell, filtered to allow the passage only of a light wavelength strongly absorbed by ozone. These detectors were also used for measurements of the temperature of the solar atmosphere.

Another experiment, taking advantage of the sunrise and sunset observed in each 90-minute polar orbit, used ion chambers that detected the absorption of the Sun's rays at these times to measure molecular oxygen density. A geocoronal experiment, using a small Lyman alpha-detecting telescope, was also included, to follow on from

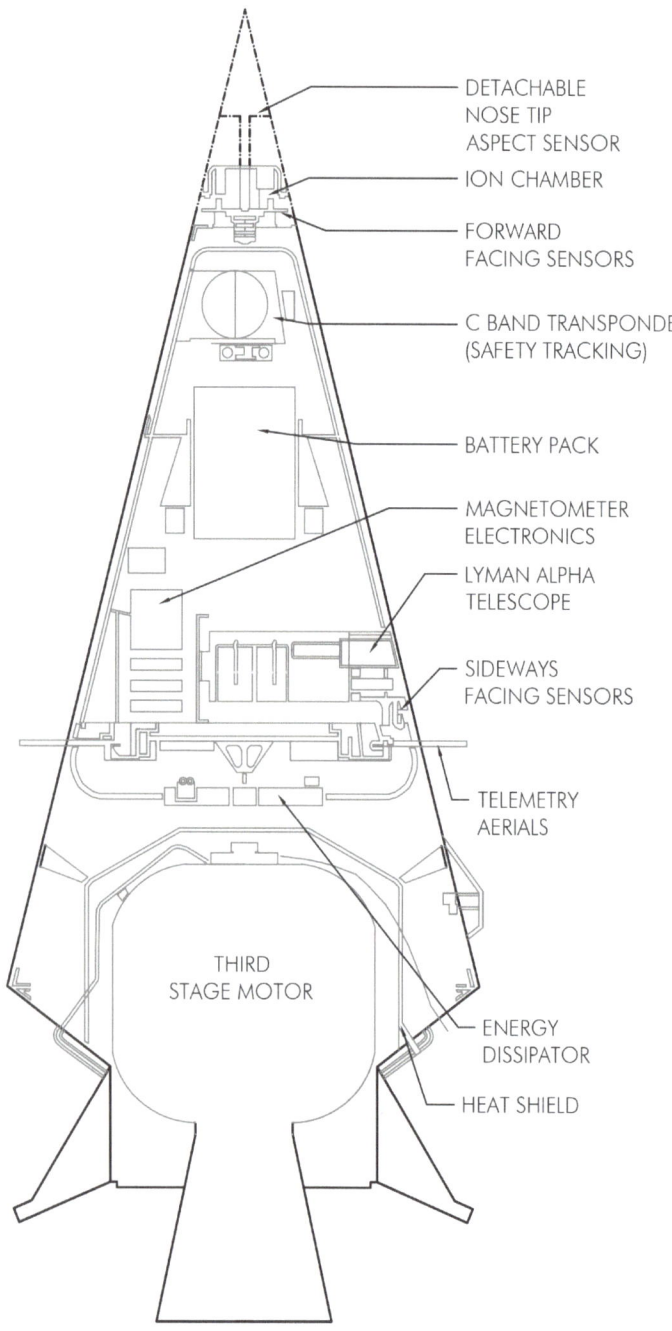

Above Cutaway diagram of WRESAT showing its main scientific instruments, based on an original WRE drawing.

TABLE 3.2 WRESAT BASIC DATA

Height: 1.59m
Weight: 45kg
Base diameter: 0.76m
Length with third-stage motor: 2.17m
Weight with third-stage motor: 72.6kg

research conducted in 1966 by a Long Tom rocket, along with a magnetometer package and two optical sensors, for accurately determining the satellite's attitude.

Given the tight development schedule, the simplest way to design WRESAT was to work within the conical shape of the existing SPARTA test nosecones and combine the satellite with the third stage motor, which would go into orbit with it. This meant that WRESAT's physical parameters and its launch weight were completely determined by the SPARTA vehicle.

Because the satellite also served as the rocket's nosecone, WRESAT would be subjected to vibration and high aerodynamic heating at launch, so an internal 'skeleton' of aluminium stringers and strong-rings was designed, to which the outer skin was attached. A heatshield of aluminium foil and glass cloth was placed between the third stage motor and the payload, to protect the scientific instruments from the heat of the motor. During launch, the sensor ports were protected by heavy duty covers that were later ejected using explosive nuts: the antennae, which were the only protrusions from the sleek nosecone satellite, were protected by cork insulation.

In order to ensure that the satellite could withstand all the rigours of its launch and orbital environments, a WRESAT prototype, fully equipped with back-up or dummy instrumentation, had to be thoroughly tested. Because suitable test equipment did not exist, WRE engineers and technicians jury-rigged the necessary test apparatus: for the torsional impact test, they found it easier to devise a rig that would decelerate the satellite from 200rpm, rather than develop a device to rapidly accelerate it to that rate; the longitudinal impact test was conducted on an improvised rig that dropped the satellite onto a block of lead.

To regulate WRESAT's internal temperature during launch and in orbit, the exterior of the satellite was painted black, with some silver highlighting, while the interior was to be painted white. On American advice, and at considerable expense, a special white paint was imported from the US. Only at the last minute was it discovered that the wrong paint had been sent and that the inside of the satellite was coated in what was effectively white household enamel! Fortunately, the temperature regulation system worked perfectly during WRESAT's short operational life.

With no time to design a solar power array, WRESAT had to be battery-powered, giving it a very limited operational lifespan. Data from WRESAT's 14 instruments and information from 15 different housekeeping functions would be transmitted back to Earth and recorded at NASA STADAN tracking stations, which could receive the data but not decode it, as only the WRE had the necessary equipment to unscramble the signals. A transportable telemetry receiving station was built to monitor the satellite's signals during launch and orbital insertion. To save time, it was improvised from components found in the WRE's warehouses, along with material borrowed from other facilities, including the NASA STADAN station at Orroral Valley, near Canberra.

For accurate orbital insertion, the SPARTA launcher's second and third stages were spin stabilised, using three small rocket motors on the second stage. However, once in orbit the satellite was designed to assume a stable end-over-end rotation that better suited the scientific instrument package. To speed up the transition from its initial spin to the required orbital attitude, an energy dissipater assembly was designed, consisting of a loop of tube containing viscous silicone oil, which dissipated the satellite's rotational energy as heat. It functioned so rapidly that WRESAT was ready to commence data gathering within minutes of entering orbit.

Because of the use of spin stabilisation, the entire launcher and satellite had to be dynamically balanced to within certain limits. Since a low-speed vertical balancing machine for the satellite was not available in Australia, the WRE developed a technique to use a commercially available horizontal dynamic balancing machine, normally used for truck engines. WRESAT was placed inside a purpose-created rig and mounted on the balance machine, with weights being added to the satellite to balance the whole assembly. So crucial

Above The WRESAT satellite under construction at WRE, Salisbury. The external paint finish and the 'kangaroo and Woomera' logo developed for the WRESAT project have been applied to the satellite's surface. (Photo courtesy of Defence Science and Technology Group)

was the dynamic balance that WRESAT was designed so faulty components could be removed and replaced without disturbing it.

WRESAT's week-long hot and cold soak, to ensure its readiness for flight, was conducted in the large vacuum chamber at the University of Adelaide. This facility had been established by Professor Carver with funds from the Australian Research Council, to facilitate the Department of Physics' upper atmosphere work, which included balloon-borne instruments and Skylark payloads as well as those for HAD and HAT. It had been designed to be large enough to contain complete

Above Liftoff! Australia's first satellite thunders on its way to orbit on 29 November 1967. (Photo courtesy of Defence Science and Technology Group)

sounding rocket nosecones and large balloon instrumentation packages and so was just capable of containing the satellite.

While WRESAT's final checks were being undertaken, its launch vehicle was being prepared by the SPARTA team at Woomera, where it would be fired from Launch Area 8. The olive-drab US Army livery of the Redstone stage was concealed with a coating of brilliant white paint that assisted with the optical tracking of the launcher against the blue sky. A special logo was created for WRESAT, consisting of a leaping kangaroo superimposed on a stylised design that symbolised both an Aboriginal woomera launching a spear and a rocket rising up from the Earth, with the seven-pointed star of the Australian flag in the background. This logo was painted on both the satellite and the launcher.

WRESAT's launch was originally scheduled for 28 November 1967 and a large party of VIPs and journalists were gathered at Woomera to witness the historic occasion. However, 30 seconds from zero a small air-conditioning unit failed to release from the rocket, causing an automatic halt to the countdown. After nine previously trouble-free launches, it was a great embarrassment to the SPARTA launch team to experience a scrub on the WRESAT launch, but the problem was fortunately minor and easily rectified. Various anecdotes about the aborted launch recount that the stuck unit was actually removed either by being kicked free or whacked with a spanner to dislodge the rust that had caused it to stick in the first place. The launch was recycled for the following day, with a six-hour countdown commencing at 8.19am. This second countdown was flawless and at 2.19pm on Wednesday, 29 November Australia's first satellite lifted off.

Two minutes after launch, the Redstone booster shut down and staging took place. The first stage fell into the Simpson Desert, where it remained undiscovered until adventurer Dick Smith located it in 1989: it was retrieved by a team of heritage-minded Woomera personnel in 1990 for display in the Woomera township. The second stage spin motors ignited at 180km, then the second stage engine fired for 30 seconds, before dropping away to fall into the Gulf of Carpentaria. Finally, WRESAT's third stage motor fired for nine seconds, placing the satellite into a 169 x 12445km elliptical orbit. The ELDO downrange tracking station at Gove in northern Australia and NASA tracking stations received signals from the satellite as it travelled around the Earth, with final confirmation that it was actually established in orbit coming from the NASA Manned Space Flight station at Carnarvon in Western Australia: it picked up WRESAT's signals on time, 99 minutes after launch.

WRESAT's battery only allowed it to operate for five days. During its 73 operational orbits, WRESAT gathered a large amount of data on upper atmospheric conditions at all latitudes, from pole to pole. Most of the instruments worked as planned. However, because of the short duration of the mission, WRESAT could do little more than provide a check on the data already gathered by HAD, HAT and other sounding rockets and could not contribute to the long-term studies of fluctuations in the levels of ozone and molecular oxygen over time.

Three days after WRESAT's launch, a Skylark rocket fitted with identical instruments was fired from Woomera. Reaching a height of 100km, against the low point of WRESAT's orbit at 169km, the two craft provided complementary data that also provided a check on WRESAT's instruments, demonstrating that they were functioning correctly. Their joint measurements showed that a 'ledge' in the zone profile at 110-120km is a common feature of the lower thermosphere, which confirmed the findings of other sounding rocket research. WRESAT's data also helped to refine the measurement of the minimum temperature of the Sun's atmosphere.

After falling silent when its batteries were exhausted, WRESAT remained in orbit until 10 January 1968, its low perigee leading to rapid orbital decay. It re-entered the atmosphere and vaporised somewhere between Ireland and Iceland.

The successful launch of WRESAT placed Australia in the company of a select group of nations that had, either totally independently, or with US assistance, launched a satellite of their own. For a country not considered by the rest of the world to be a major technological power, it was a notable technical and scientific achievement and clearly demonstrated the skills of Australian science and industry.

Australia garnered congratulations and praise from the international community: even the USSR expressed a welcome to the 'Space Club', while

RIGHT Technicians installing a scientific instrument in WRESAT in a laboratory at WRE Salisbury. (Photo courtesy of Defence Science and Technology Group)

AUSTRALIA'S PLACE IN THE SPACE CLUB

It is often claimed that Australia was either the third or fourth country to launch its own satellite, but the exact order of precedence is, in fact, a more complex matter. France had developed its own independent launch vehicle and orbited its first satellite in 1965 from a base in Algeria, which had previously been a French colonial territory. Canada had built a satellite that was launched for it in the United States in 1962. In the same year, the UK provided the instruments for a 'British' satellite built and launched by NASA. An Italian built satellite was launched by the US in 1964, and a second Italian satellite was launched, using an American-supplied Scout rocket, in April 1967, from Italy's San Marco mobile launch platform, based off Kenya.

This author believes that the most appropriate sequence is for France to be recognised as third after the USSR and United States, since it independently developed both its rocket and satellite, with Italy fourth and Australia fifth, since both nations used American rockets to launch their own satellites from national launch facilities.

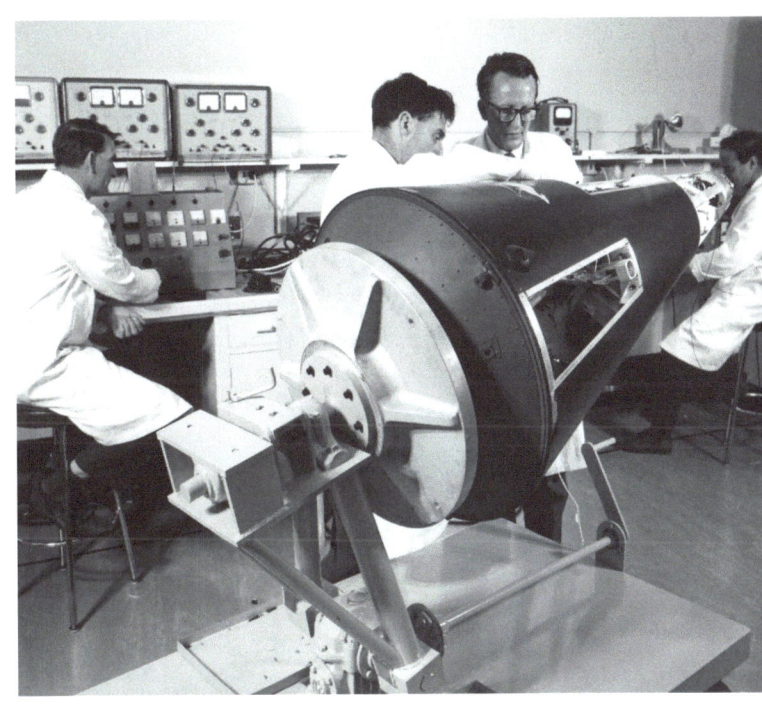

US President Johnson was quoted as saying that WRESAT 'shines as brightly as the Southern Cross'. The Australian media, initially dismissive after the first aborted launch attempt, heaped euphoric praise on the satellite project after the launch and Australian politicians, despite their lukewarm support for WRESAT when it was first proposed, were quick to bestow and accept congratulations for the achievement. In February 1968, Fairchild Australia presented the WRESAT team with its Planar Award for outstanding achievements in the Australian electronics industry. A WRESAT display, initially mounted at Parliament House in Canberra, toured the state capital cities and was then presented internationally at conferences and trade fairs as a demonstration of Australian technological prowess.

All those involved in WRESAT had hoped that a successful first mission might lead to funding for further satellites: it was for this reason that the official designation of the satellite was actually WRESAT 1. Carver's team at Adelaide University were particularly interested in establishing a database of scientific readings obtained from orbit over long periods, to complement the sounding rocket data. However, WRESAT 1 was never to be followed by WRESAT 2. Despite the United States offering to provide a number of Redstone launchers at greatly discounted prices, there was no interest or commitment within the Australian Government to establishing an Australian 'space presence'. Cabinet had been convinced to support the satellite project on the basis that it represented a cheap way for Australia to achieve international prestige as a 'player' in the Space Age. However, the politicians' vision did not extend beyond the international cachet of having achieved the launch of a national satellite.

The WRESAT Project enabled Australia, by taking advantage of a generous gesture from the United States, to become one of the earliest nations to build and launch its own satellite. Although WRESAT's scientific contributions were negligible, it was a significant technical achievement and demonstrated the WRE's level of engineering expertise in being able to design and construct a satellite in so short a time. The fact that WRESAT had no successors was not due to any failure on the part of its technical and scientific development teams, but to government disinterest and an inability to perceive the long-term benefits to Australia that could have accrued from a home-grown satellite program built upon the success of WRESAT 1.

LEFT *In September 2017 Australia Post released this commemorative stamp and a variety of philatelic collectables to celebratie the 50th anniversary of WRESAT's launch. (Image courtesy of Australia Postal Corporation, © Australian Postal Corporation, 2017)*

Australis-OSCAR 5:
Australia's First Student Satellite

WRESAT was not the only Australian satellite to be constructed in the 1960s: in fact, a group of undergraduate students at Melbourne University began constructing their satellite before WRESAT was even on the drawing board! Although Australis-OSCAR 5 had to wait several years for a launch by NASA in the United States, it became Australia's second satellite.

As early as 1962, a group of American amateur radio enthusiasts had constructed the first Orbiting Satellite Carrying Amateur Radio (OSCAR). Their early success encouraged members of the Melbourne University Astronautical Society (MUAS) to build their own amateur radio satellite, the first to be constructed outside the US.

Formed in the mid-1960s, MUAS members set up equipment to track and receive signals from various American and Soviet satellites. Several amateur groups in Australia were tracking satellites at that time, but the MUAS students, with support and assistance of university staff, were the only ones to receive images on a regular basis. One of their achievements was the first regular reception in Australia of meteorological images from early American TIROS and Nimbus weather satellites. They supplied visible and infra-red weather images daily to the Bureau of Meteorology before it established its own receiving facilities and were even able to assist NASA in identifying a fault that had occurred in a Nimbus satellite, because of a transmission that they received.

MUAS also tracked the early OSCAR amateur radio satellites, which were launched by the US Air Force, piggy-backing on major satellite launches. These small satellites enabled radio amateurs to gain experience in satellite tracking and conduct experiments in radio wave transmission through the atmosphere. After receiving signals from OSCAR 3 and 4, the Melbourne group was inspired to consider, in conjunction with the Melbourne University Radio Club, an Australian amateur radio satellite.

The students decided to build a small 'beacon' satellite, which, like a radio beacon, would transmit telemetry data back to Earth on fixed frequencies. Such a limited technological goal proved to be a wise decision in view of the jungle of official approvals that was later found to be necessary, even for this simple satellite.

The satellite project, which the students called Australis, began in March 1966. Volunteers from MUAS, other university societies and university staff worked together to get Australis off the ground. With limited funding, electronic and other components were mainly acquired by donation from suppliers, some of whom even based their own advertising on their donations. Technical and financial assistance was also received from the Wireless Institute of Australia, although the Australis team paid most expenses out of their own pockets.

BELOW Student-built amateur radio satellite, Australis-OSCAR 5. Look closely and you'll see that the antennae were cut from a flexible steel measuring tape. (Photo courtesy of Owen Mace)

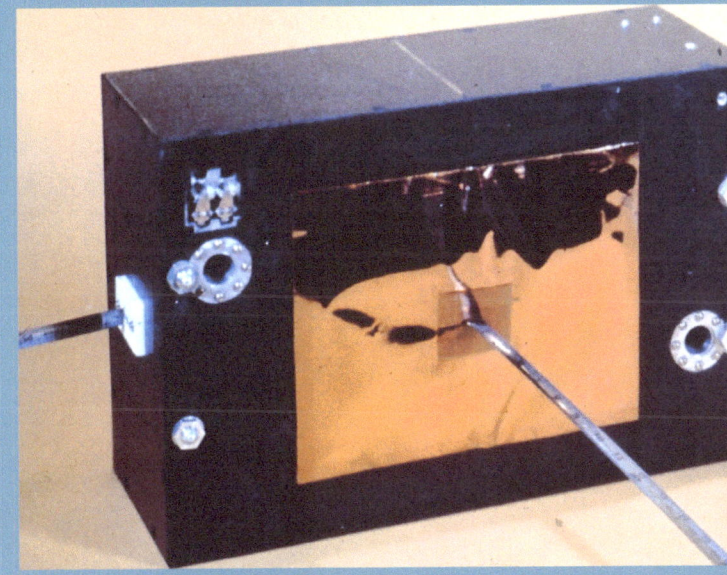

Australis-OSCAR 5: Australia's First Student Satellite

With a pace of development equal to that of WRESAT, in little more than a year the satellite went from drawing board to completion: in fact, Australis was actually completed and ready to fly before WRESAT was even commenced. Like WRESAT, Australis was a classic example of 'making do', and every step of the way was a learning experience for the students, some of whom went on to become professionally involved in Australian space activities.

University facilities were pressed into service to assist during the construction and testing of the satellite, with the Post Master General's Department (now split between Australia Post and Telstra) and the Department of Supply also providing testing facilities that the university did not have available. A number of components, including the transmitters and command system, were tested by being flown on high altitude research balloon flights to ensure that they were working correctly and were robust enough to withstand the rigours of spaceflight. Permission also had to be sought for the transmitters to use certain frequencies. The NASA representative in Australia gave the project significant assistance in arranging matters in the United States, especially when the satellite's launch was finally allocated to a NASA weather satellite launch.

The completed satellite was delivered to the San Francisco headquarters of Project OSCAR in June 1967, but there were many delays in finding a launch for it to hitch a ride into space. With the formation of AMSAT (Radio Amateur Satellite Corporation), the satellite was shipped to its headquarters in Washington DC in March 1969. It took another nine months for AMSAT to put the satellite into final shape so that it would be approved by NASA as suitable for launch; to arrange with NASA for the launch itself; and to wait out the many agonising delays caused by booster difficulties. NASA required that Australis have some scientific or technological merit, and AMSAT justified the launch on the basis that the satellite would provide experience and training for radio amateurs and allow for the investigation of unusual radio signal transmission through the Earth's atmosphere and ionosphere.

After long delays, the first non-American OSCAR satellite was launched from Vandenberg Air Force Base at 9.31pm (EAST) on 23 January 1970 by a Thor-Delta rocket carrying, as its main payload, the TIROS-M (ITOS-1) weather satellite. The satellite was placed into a 115-minute orbit, varying in altitude between 1416-1464 kilometres. Once in orbit, the satellite was called Australis-OSCAR 5, or simply AO 5. It was the first Project OSCAR satellite launched since 1966 and the first by NASA for one of its satellites.

Weighing only 39 pounds (17.7 kilograms) the satellite was constructed in the form of two cubic aluminium shells. It carried two small transmitters, each beaming the same telemetry signal on 29.450 MHz in the 10-metre band and 144.050 MHz in the two-metre band. The telemetry system was sophisticated, but designed for simple decoding without expensive equipment. The start of a telemetry sequence was indicated by transmitting the letters HI in Morse code. Seven parameters were then transmitted, providing data on battery voltage, current, and the temperature of the satellite at two points as well as information on the satellite's orientation in space from three horizon sensors. Power was supplied by two strings of chemical batteries and the electronics were mounted in a frame built around the batteries.

AMSAT applied an experimental paint pattern to the outer skin, to control the internal temperature of the satellite, but temperature data indicated that the satellite was operating with an internal temperature of around 50°C. Fortunately, Australis' instruments were robust and had been designed to function at high temperatures. The antennae were made of flexible steel tape: instead of being folded, to be extended once in orbit, they were attached at one end and wrapped around the satellite in the launch vehicle to await deployment.

When Australis was released from its launcher, springs pushed it away into its own orbit. These were made by a mattress manufacturer in Melbourne who produced a pair of springs carefully matched in force, so that the satellite would not spin when released. At the same time, the electronics were turned on, while the steel tape antennae unwound themselves. A passive magnetic attitude stabilisation system, which maintained the satellite's orientation by reference to the Earth's magnetic field, was used for the first time in an amateur satellite to stabilise Australis and reduce its rate of spin. This system helped to reduce signal fading, making its signals easier for ham radio operators to detect the satellite's transmissions.

A special technique was devised to permit any station in the world to track the satellite. The first radio amateur to report receiving the beacon transmission was on the island of Madagascar, who detected the two-metre signal 66 minutes after launch. Another amateur in Darwin reported reception of the 10-metre signal a few minutes later. Other radio amateurs in Western Europe and North America also reported receiving both the two- and 10-metre signals on the satellite's first orbit. The use of the 10-metre band was also a first for AO 5, as it had not previously been used for satellite transmission. During its initial orbits AO 5 passed within range of Melbourne and members of the Melbourne University Radio Club were able to tune in to the satellite. At the end of the first orbit, however, problems developed with the 10-metre transmission, which eventually became very difficult to decode.

With each successive orbit, telemetry calibrations, technical data, orbital information and predictions, and reception reports were collated by amateur radio operators. By the end of Australis' first day of operation, AMSAT headquarters had already received more than 100 tracking, telemetry and reception reports. Reports were eventually received from several hundred stations in more than 27 countries. Several radio clubs performed extensive tracking and telemetry data recording, with the observations recorded on standard reporting forms that were suitable for computer analysis.

The two-metre signal was broadcast continually and operated until 14 February. The 10-metre transmission continued until the satellite's batteries failed around 9 March. To conserve the battery power, stations in Australia and the US used a prearranged schedule to switch the transmitter on (mainly on weekends) and off (during the week) via the first successful command system installed in an amateur satellite. The demonstration of command capabilities was to prove very important in obtaining US Federal Communications Commission licences for future amateur radio satellite missions. Performance measurements of the 10-metre beacon confirmed that this band would prove suitable for transponder downlinks on future spacecraft, and led to its use on OSCARs 6, 7 and 8. Although its batteries eventually failed, Australis-OSCAR 5 remains in orbit and will continue to do so for several hundred years.

Following the success of AO 5, the Project Australis group provided equipment for a couple of subsequent AMSAT satellites. They also considered building a solar-powered multi-channel repeating satellite, which would both transmit and receive signals. However, due to the difficulty of finding funding and the need for the students to concentrate more on their studies and family life, this and similarly ambitious projects were to remain unrealised.

Although Australia has not launched another amateur radio satellite, amateur radio enthusiasts here have continued to use satellites and today AMSAT Australia operates as a division of the Wireless Institute of Australia, co-ordinating national amateur satellite activity. The group boasts several hundred members who utilise a fleet of amateur radio satellites built by teams from many nations.

BELOW Student satellite builder Owen Mace, holding a model of Australis-OSCAR 5 showing the experimental temperature control paint pattern applied by AMSAT before launch in the US. (Photo courtesy of Owen Mace)

Above Pre-flight portrait of America's first Earth-orbiting astronaut, John Glenn. The inclusion of a tracking antenna at Cape Canaveral emphasised the importance of space tracking to the success of the astronaut's mission. (Photo courtesy of NASA)

Right The historic Honeysuckle Creek antenna, which served in both the Manned Spaceflight and Deep Space Networks, in operation at Tidbinbilla. (Photo courtesy of NASA/CSIRO)

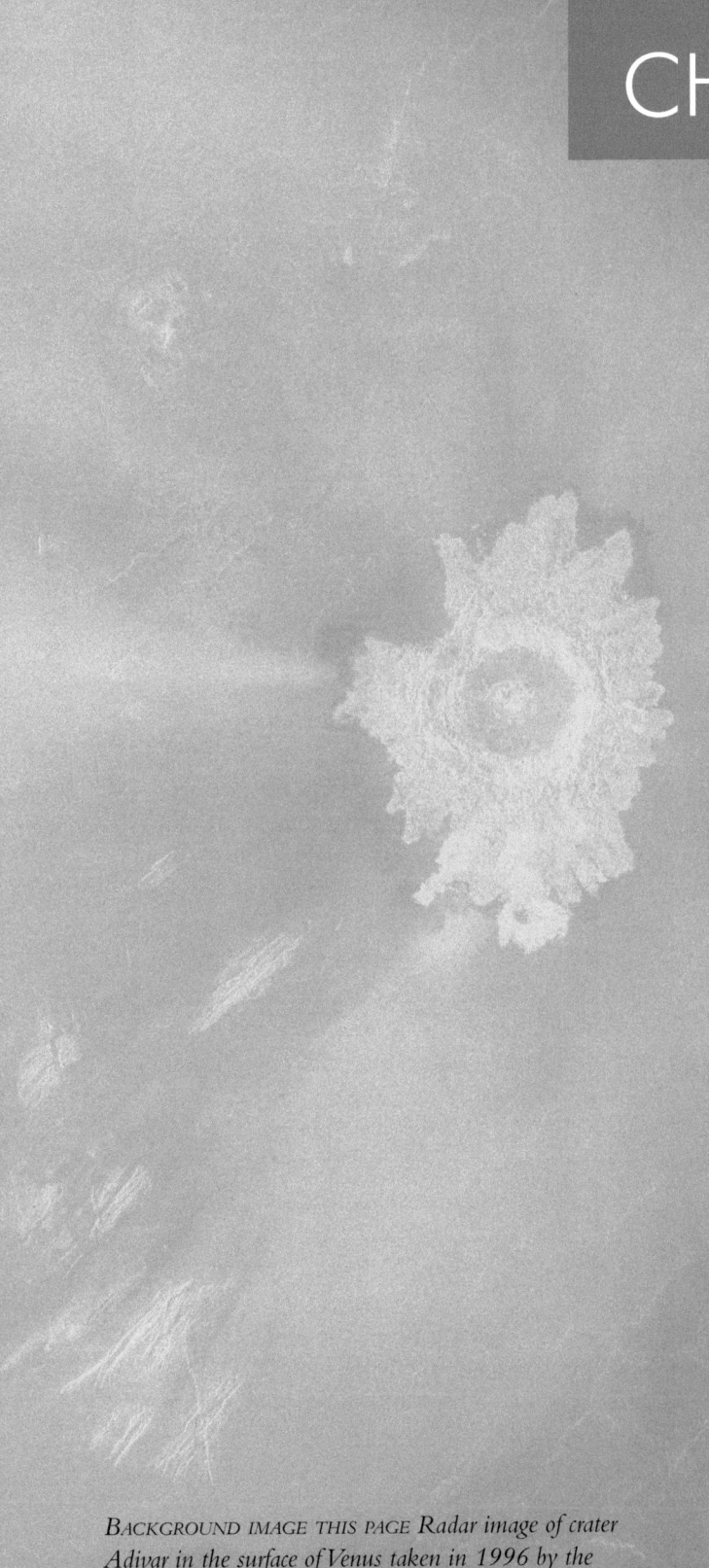

BACKGROUND IMAGE THIS PAGE *Radar image of crater Adivar in the surface of Venus taken in 1996 by the Magellan spacecraft, one of the many planetary exploration missions supported by NASA's Deep Space Network station at Tidbinbilla. (Image courtesy of NASA/JPL)*

CHAPTER 4

Keeping on Track: Space Tracking From Australia

The establishment of the US Minitrack and SAO Baker-Nunn tracking facilities in Australia for the International Geophysical Year marked the beginning of Australia's longest continuous space activity: providing space tracking and communications services for the orbital and deep space programs of the United States and other nations.

Between 1960 and the 1980s, Australia would host facilities supporting all three of NASA's tracking networks: the Deep Space Network (DSN – for planetary exploration missions); the Space Tracking and Data Acquisition Network (STADAN – for Earth-orbiting satellites) and the Manned Space Flight Network (MSFN – for human spaceflight missions). The country would, for a time, become the home of the largest number of NASA space tracking facilities outside the United States, enabling Australian participation in many significant space exploration missions. The tradition of space tracking involvement still continues today with NASA's Deep Space Communication Complex outside Canberra and the ESA tracking station at New Norcia in Western Australia.

Australia Joins the NASA Network

As outlined in Chapter 2, the IGY Minitrack and SAO Baker-Nunn facilities were originally established under the control of the Naval Research Laboratory and the Smithsonian Astrophysical Observatory, since the National Aeronautics and Space Administration (NASA) did not exist at that time. When NASA came into being in October 1958, the new agency immediately became responsible for many space projects under development by the various branches of the US Armed Services that were primarily scientific and civilian in nature, including the IGY tracking networks.

To track and monitor the new Earth-orbiting satellites that it planned to launch, NASA required a global tracking network and modified the Minitrack stations to become the nucleus of its Space Tracking and Data Acquisition Network. NASA also assumed responsibility for the day-to-day operations and partial funding of the SAO Baker-Nunn facilities, although their observation program and the co-ordination of the Moonwatch groups, which also continued to operate after the IGY, remained with the Smithsonian Astrophysical Observatory.

Plans for lunar and planetary exploration previously commenced by the US Department of Defence and the Army's Jet Propulsion Laboratory (JPL) were transferred to NASA, along with JPL itself. These missions to the Moon and planets required the development of a second tracking network specifically designed for the unique needs of deep space missions beyond Earth orbit. Because the Earth is rotating, deep space tracking networks require a minimum of three stations, each located 120° apart: as the planet turns, at least one station will always be facing in the direction from which the spacecraft's signals are coming. Tracking stations needed to be located in both the northern and southern hemispheres, to provide coverage of the entire sky, since a space probe's path, as seen from Earth, can cross both hemispheres.

In late 1958, when NASA announced its first human spaceflight program, Project Mercury, the mission safety requirements of a piloted spaceflight led to the establishment of a dedicated tracking and communication network that could provide real-time voice communications with the astronaut, medical and spacecraft status telemetry, and a radio command system for activation of the re-entry system. Initially named the Mercury Space Flight Network, this third network was managed, like STADAN, from the Goddard Space Flight Centre, near Washington DC. When the Apollo lunar landing project began, the network was greatly expanded and re-named the Manned Space Flight Network

Australia's geographic position with respect to the United States made it an ideal location for all three of NASA's

Left DSIF 41, located at Island Lagoon, near Woomera, was NASA's first deep space tracking station outside the United States. This dramatically illuminated night view shows the station's antenna in its later S-band configuration. (Photo courtesy of NASA)

planned networks. Because the first orbit of almost all spacecraft launched from Cape Canaveral would pass within sight of Western Australia, locating tracking facilities there would enable the confirmation that a spacecraft was safely in orbit. Also, in order to maintain an unbroken network of contact with a satellite or crewed spacecraft, tracking stations would be needed between Africa and Hawaii and, because the continent occupies such a broad swath of the Earth's surface, Australia was well placed to provide the location for one or more STADAN and MSFN stations. Being located 120^0 around the world from JPL, which still today manages the Deep Space Network from California, also meant that Australia was perfectly placed to host a station to support lunar and planetary exploration missions.

The Space Co-operation Agreement: An Enduring Partnership

The original IGY tracking stations were each operated under separate agreements between Australia and their US parent bodies. They were managed by the Department of Supply through its agency, the Weapons Research Establishment. Under these agreements, Australia insisted that the facilities should be managed by Australian staff and operated without direct interference from the United States. In return, the Australian Government donated the land, buildings, and technical and scientific staff for the construction of the stations, as a gesture of support for its strategic ally.

However, with its expanding plans for deep space exploration, human spaceflight and Earth-orbiting satellites, NASA needed to establish a new umbrella agreement that would cover all its future operations in Australia. Consequently, on 26 February 1960, the Governments of Australia and the United States formally agreed to co-operate in spacecraft tracking and communications through the Space Co-operation Agreement. In this treaty, NASA and Australia jointly established a management policy, which has proved very successful, allowing the agreement to continue, with very little change, up to the present day.

Following the precedents established with the IGY stations, NASA would finance the construction and operation of its tracking stations and be responsible for system design and policy formulation. Australia was responsible for the detailed facilities design and the installation, operation and maintenance of the stations, as well as for staffing the facilities. Because Australian companies undertook all the construction, operations and maintenance work, local industry was able to develop and maintain considerable expertise in areas of electronic and mechanical engineering, such as antenna design, relevant to both the tracking stations and the satellite communications ground sector. Under the agreement, Australia also had the right to use the stations for its own scientific research, provided such research did not conflict with NASA's priority tasks.

Table 4.1 Major NASA Tracking Facilities in Australia

Station	Location	Years of Operation		
Minitrack	Woomera, SA	1957-1966	Orroral Valley	1966-1985
Baker-Nunn	Woomera, SA	1958-1975	Orroral Valley	1975-1982
Muchea	*Perth, WA*	*1960-1964*		
Red Lake	*Woomera, SA*	*1960-1966*		
Island Lagoon	**Woomera, SA**	**1960-1972**		
Carnarvon	*Carnarvon, WA*	*1963-1975*		
Tidbinbilla	**Tidbinbilla, ACT**	**1964-present**		
Orroral Valley	Orroral Valley, ACT	1965-1985		
Honeysuckle Creek	***Honeysuckle Creek***	***1967-1981***		

Deep space tracking facilities indicated in **bold**.
Manned spaceflight stations indicated in *italics*.
Honeysuckle Creek was both a MSFN and a DSN station across its lifetime. Carnarvon also acted as a STADAN facility.

Initially, the WRE managed and staffed the stations, but in 1962, when the extent of the tracking station projects and the potential benefits in technology transfer became apparent, the WRE established a special section to streamline station management, while the day-to-day operation of the tracking stations was contracted to private industry, a practice which still continues today. At their peak, during the 1960s and 1970s, more than 700 staff were employed at NASA tracking stations. Many personnel transferred to the tracking stations from the WRE and various programs at Woomera. In the 1960s, when qualified local technical staff were not always available, skilled British technicians were encouraged to migrate to Australia and take up positions at the tracking stations.

Management of the NASA facilities was transferred to the Department of Science when the Department of Supply was dissolved in 1975, and then passed to the newly-created Australian Space Office in the late 1980s. When the Space Office was terminated in 1996, management of the last remaining major NASA station, the Canberra Deep Space Communication Complex at Tidbinbilla was passed to the CSIRO, which still oversees the management of the station through its Astronomy and Space Science division. To oversee its interests in Australia, NASA also established a liaison office in Canberra in 1962, employing a Senior Scientific Representative, with representatives of the Jet Propulsion Laboratory and the Goddard Space Flight Centre overseeing the specific networks for which their NASA Centres were responsible.

During space missions, each tracking station communicated directly with the mission control centre in the US by teletype and voice circuits, to report on the position of a spacecraft and to pass on the latest engineering and scientific data received. In Australia, OTC, with assistance from the Postmaster General's Department (PMG), provided and maintained ground-based communication links via telephone lines, undersea cables and radio, as well as international telecommunications satellite links. Special NASCOM (NASA Communications) switching centres were established, first in Adelaide and later at Deakin in Canberra, to ensure reliable, priority communications between the United States and Australia during space missions. Following the launch of the first satellite in NASA's Tracking and Data Relay Satellite System (TDRSS), in 1983, the ground-based links were gradually phased out.

BELOW NASA's second DSN station at Tidbinbilla, as it looked in 1965. Today, as the Canberra Deep Space Communication Complex, Tidbinbilla is the major NASA facility in Australia – a testimony to the strength of the Space Co-operation Agreement, the skills of Australian industry, the dedication of its technical staff and its particular geographic location. (Photo courtesy of NASA/CSIRO)

THE SATELLITE TRACKING AND DATA ACQUISITION NETWORK (STADAN): SUPPORTING RESEARCH IN EARTH ORBIT

When NASA acquired control of the Minitrack network, it became the basis of its Satellite Tracking and Data Acquisition Network (STADAN; from 1975 called the Spaceflight Tracking and Data Network, or STDN), primarily intended for control and data reception of its Earth-orbiting satellites. The Minitrack station at Woomera's Range G thus became Australia's first STADAN facility. However, NASA planned that its next generation of Earth-orbiting satellites would transmit at a frequency of 136MHz, rather than the original 108MHz used by the first US satellites. In 1961, it took the opportunity not only to upgrade the Woomera Minitrack station, but also to move it to the site of its Deep Space Instrumentation Facility at Island Lagoon (see below), to consolidate NASA activities at Woomera. The SAO Baker-Nunn telescope was also moved to Island Lagoon in 1963.

The workhorse STADAN stations provided support for NASA's growing armada of Earth orbiting scientific and experimental applications satellites and it was soon obvious that an additional station would be needed in Australia. Due to the cost and difficulties of maintaining station staff at the remote Woomera location, NASA decided to establish its second generation of Australian tracking stations in the Australian Capital Territory and selected a site in Orroral Valley, about 60km south-west of Canberra by road.

The new Orroral Valley station's main antenna was a 26m (85ft) steerable dish, which was joined in 1967 by the Minitrack equipment from Island Lagoon. The Baker-Nunn satellite camera would also be moved from Island Lagoon to Orroral in 1975, following the closure of the deep space station there. Across its lifetime, the Orroral Valley station supported not only an array of Earth-orbiting spacecraft, but also received data from the Apollo Lunar Surface Experiment Packages (ALSEPs) left on the Moon by each Apollo landing mission. It later provided tracking and communication with the Apollo-Soyuz Test Project (the first space link-up between the USSR and the US, in 1975) and early Shuttle missions.

With the development of NASA's TDRSS satellite network, Orroral Valley was closed in 1985. The Tidbinbilla DSN station then took over the remaining tracking tasks for Earth-orbiting satellites, using the 26-metre antenna which had been transferred from Honeysuckle Creek. The original Orroral Valley 26-metre antenna was moved to the University of Tasmania for use as a radio astronomy instrument. The SAO Baker-Nunn telescope was donated to the University of New South Wales in 1982, where it was extensively refurbished to become the university's Automated Patrol Telescope. Located since 1994 at the Siding Spring observatory complex near Coonabarabran, NSW, it has played an important role in the university's extra-solar planet search program.

Above Orroral Valley tracking station in the 1970s. The flat Minitrack antenna array from Island Lagoon is in the foreground. (Photo courtesy of www.honeysucklecreek.net)

Left NASA's first Tracking and Data Relay Satellite (TDRS) launched from the Space Shuttle Challenger in 1983. The deployment of the TDRS network reduced the need for NASA to rely on overseas tracking stations and led to the closure of most of its facilities in Australia. (Photo courtesy of NASA)

The Deep Space Network: Exploring the Solar System by Proxy

Although humanity has not yet explored beyond the Moon in person, from the very beginning of the Space Age attempts were made to send space probes out to discover more about the Moon and then the planets beyond. These missions, carried out with increasingly sophisticated automated spacecraft, have allowed humanity to explore the Solar System by proxy: the data and breathtaking images they have sent back to Earth have been received at the stations of the Deep Space Network, the second NASA tracking network to be established.

Island Lagoon Tracking Station

The first DSN tracking station outside the United States was Deep Space Instrumentation Facility (DSIF) 41: in this designation the 4 represents the country (Australia) and the 1 means that it was the first station built in that country. Constructed in 1960 on the edge of the Island Lagoon salt lake, about 27 kilometres south of the Woomera township, the station was equipped with a 26-metre polar-mounted antenna. Its design was based on the dish already constructed at NASA's deep space tracking station at Goldstone, California. The Goldstone antenna design itself was based on a radio telescope design, which was at least in part the work of John Bolton, the first Director of the Parkes Radio Telescope.

Officially opened in April 1961 by receiving signals from Goldstone, bounced off the Moon, Island Lagoon originally operated in the L-band, but was converted to S-band in 1964 to support more advanced spacecraft. Although many of the early missions the network tracked were unsuccessful, the performance of the tracking stations themselves demonstrated the importance of a world-wide network controlled from a central location (in this case JPL), as an essential element in future space missions to the Moon and planets. DSIF 41 played a crucial role in NASA's first successful planetary mission, Mariner 2 to Venus in 1962, entrusted with the responsibility of commanding the spacecraft to manoeuvre so as to point its high gain antenna continuously towards the Earth. During the critical period (roughly 40 minutes) when Mariner 2 was passing Venus, Island Lagoon was the only station to maintain continuous contact with the spacecraft, obtaining good quality data reception from the spacecraft's 3-watt transmitters.

Following this success, Island Lagoon (now identified as DSS 41, following the reorganisation and re-naming of the Deep Space Network in late 1963-64) went on to provide support for the later missions of the Ranger series of Moon probes and joined the new Tidbinbilla station in supporting Mariner 4, the first successful mission to Mars. In conjunction with Tidbinbilla, Island Lagoon supported the remaining single-planet missions in the Mariner series: Mariner 5 to Venus and Mariners 6, 7 and 9 to Mars. The two stations also worked together to provide long term tracking and data acquisition support for the solar orbiting Pioneer missions, investigating the interplanetary space environment. In the last phases of automated lunar exploration, prior to the Apollo missions, Island Lagoon supported the Lunar Orbiter program, designed to take high resolution photographs of the lunar surface to aid in the selection of suitable landing sites for the Apollo astronauts.

DSS 41 would continue in operation until 1972, when it was closed for economic reasons: the final mission it supported was Pioneer 10, then on its way to Jupiter. Following the closure, its electronic equipment was transferred to the Honeysuckle Creek MSFN station near Canberra. Although the tracking antenna was offered to Australia as a potential radio astronomy instrument, it was eventually deemed that it would be uneconomical to transport it to another location and the dish was sold for scrap.

Tidbinbilla Tracking Station

With ambitious plans for lunar and planetary exploration in the period 1965-68, NASA realised that it would need additional stations to expand its mission support capacity and embarked upon a program to construct a 'second network' of DSN stations, one of which was to be built in Spain (lying roughly along the same longitude as the existing South African DSN station), the other in Australia. As with the expansion of the STADAN network, NASA decided that it would establish its second Australian DSN station close to Canberra, and selected a site in the Tidbinbilla Valley, about 40km from Canberra. The development of Tidbinbilla was fast-tracked as much as possible, so that the station would be available to support the Mariner 4 Mars mission in 1965.

A team of Australian engineers and technicians was based for the first half of 1964 at the Goldstone tracking station, to become familiar with the Deep Space Network and assist in the assembly and testing of the electronic equipment designed for the Australian station. This same team then carried out the installation and commissioning of the equipment at Tidbinbilla itself. This co-operative exercise worked so well that it became the model for most future development/upgrade projects of DSN facilities in Australia and elsewhere.

DSS (Deep Space Station) 42, Tidbinbilla, was equipped with a 26-metre polar-mounted antenna like that at Island Lagoon, although it utilised

Left NASA illustration of the Mariner 2 spacecraft, the first planetary mission to successfully reach another world and return data. The Island Lagoon tracking station played a crucial role in tracking Mariner 2 and receiving the data it sent back to Earth. (Photo courtesy of NASA)

Above Dubbed the 'picture of the year' and even the 'picture of the century' by the media in 1966, this oblique view across the lunar crater Copernicus was taken with a telephoto lens by Lunar Orbiter 2, a mission supported by Island Lagoon. (Photo courtesy of the Lunar and Planetary Institute)

newer technology in its signal amplifier and operated only in the S-band. Copying the practice at Goldstone, where the various antennae were given names of their own, the original Tidbinbilla dish was named 'Weemala', from an Aboriginal word meaning 'a distant view'. Although station staff lived in Canberra and commuted the 40km to the station each day, a personnel building at the station provided catering facilities and limited emergency sleeping accommodation, in the event that a mission crisis required staff to be on hand for long hours.

Tidbinbilla commenced operations in December 1964, to support the Mariner 4 mission, but its first space probe launch was that of the solar-orbiting Pioneer 6, which carried a cosmic ray detector designed by Australian physicist Dr Ken McCracken. Nervous about supporting their first launch, the trackers at Tidbinbilla asked so many questions of their colleagues at JPL that their concerns helped to established the standard contingency procedures for the DSN.

As Mariner 4 approached Mars in July 1965, Tidbinbilla and Island Lagoon worked together to support its planetary flyby. Because the signal from Mars was so weak when it arrived back on Earth it was difficult to receive, so Tidbinbilla requested that all aircraft be diverted that might come between the station's antenna and the signal from Mars. This resulted in a humorous incident: just at the time when Mariner 4 passed behind Mars, DSS 42 received its first ever call on its special direct line from Canberra Airport, asking if they were experiencing interference from a UFO! Later the object was identified as a weather balloon.

In conjunction with DSS 41, Tidbinbilla supported the later Mariner missions to Mars and Venus and the Pioneer missions. However, while Island Lagoon focused on the Lunar Orbiter Apollo precursor missions, Tidbinbilla supported the Surveyor lunar soft-landing program, which would demonstrate that it was safe for an Apollo Lunar Module to land on the Moon. A highlight of Tidbinbilla's involvement in the Surveyor program was the 'awakening' of Surveyor 1 after its first lunar night. The solar powered Surveyor spacecraft could not function during the long lunar night, so a successful return to operation after a 14-day 'cold soak' was vital to the success of the ongoing mission. For his successful handling of the first Surveyor awakening, one Tidbinbilla shift supervisor received the 'Prince Charming Award' from the network.

As will be described below, during the Apollo program itself Tidbinbilla worked in conjunction with the nearby Honeysuckle Creek MSFN station to provide support to the two components of the Apollo spacecraft—the Command Module and the Lunar Module—which would operate independently during their time at the Moon.

Above Left In November 1969, the crew of Apollo 12 visited the Surveyor 3 lander, which had arrived on the Moon in 1967. Both missions were supported by the Tidbinbilla DSN station. (Photo courtesy of NASA)

Above Right This image showing a Martian crater was one of the 22 pictures taken by Mariner-4, the first successful Mars mission. Mariner 4 was the first planetary mission supported by the Tidbinbilla DSN station. (Photo courtesy of NASA/JPL)

As the deep space missions for the 1970s would venture further and further into the Solar System, NASA needed larger antennae to receive the ever-fainter signals that would be sent back to Earth. It would also be necessary to transmit very powerful radio signals in order to send commands to distant space probes. Consequently, in 1969 construction work commenced on Tidbinbilla's second antenna, a 64-metre dish whose design was based upon that of the Parkes Radio Telescope (also discussed further below). Designated DSS-43, and named 'Ballima', meaning 'very far away', this antenna became partly operational in late 1972, supporting Apollo 17 as its first mission. Designed to support the Viking Mars landing missions (1976), the 64-metre dish was 3.5 times more sensitive than the 26-metre antenna. However, by the 1980s, still greater sensitivity was required to support the Voyager missions to Uranus and Neptune and both the dishes were further enlarged: 'Weemala' to 34 metres and 'Ballima' to 70 metres.

ABOVE Viking 1 view of the surface of Mars. The command that sent the lander on the way to its historic landing was issued from Tidbinbilla. (Photo courtesy of NASA)

BELOW Voyager 2 visited all the gas giant planets of the outer Solar System in the 1980s. Tidbinbilla has tracked Voyager 2 since its launch, to the edge of the Solar System and beyond. (Photo courtesy of NASA)

Over the decades, the Canberra Deep Space Communications Complex (CDSCC) at Tidbinbilla has continued to grow, with regular upgrades and improvements made to its antennae and equipment. In 1982, the antenna from the former Honeysuckle Creek MSFN station was moved to Tidnibilla, where, under the designation DSS 46, it took over the tracking of Earth orbiting spacecraft in preparation for the closure of the Orroral Valley STDN station. It remained in operation until 2009. An additional 34-metre dish, DSS 45, was added to the station in 1986, to help support the Voyager 2 spacecraft's encounter with the planet Uranus: it remained in use until 2016.

As the number of deep space exploration craft increased, as well the distances from the Earth that some of them were reaching, 1997 saw the installation of DSS 34, the first of a new generation of 34-metre antennae using beam wave guide technology. In 2014, DSS 35 became operational to assist in the tracking of space probes from many other nations and space agencies, and most recently the 34-metre DSS 36 entered service in 2016. The original Tidbinbilla dish, DSS 42, was retired in the year 2000.

Since the 1970s, Tidbinbilla has supported every NASA space mission venturing beyond Earth orbit, often playing a crucial role at critical points in the mission. For the Viking missions of 1976, for example, DSS 43 would transmit the 'Go' command that would separate the Viking 1 Lander from its Orbiter and send it on its way to providing the first striking images from the surface of Mars. More recently, CDSCC provided the prime contact and relay with NASA's Curiosity rover as it made its way to a safe landing in Gale Crater on Mars in 2012 and then received its first images and data. In 2014, Tidbinbilla continued its program of co-operative support for deep space missions from non-NASA nations, begun in the 1980s, by monitoring the orbital insertion around Mars of India's first planetary probe, Mangalyaan – an achievement which made India the first country to successfully reach Mars orbit on its first attempt.

When NASA's New Horizons spacecraft flew past Pluto in 2015, Tidbinbilla provided the final contact before the encounter and then received the first close-up images of Pluto and its surface. It also participated in a mission critical radio experiment, beaming signals to a New Horizons instrument to determine the structure of Pluto's atmosphere. In 2017, CDSCC will receive the final

Above Canberra Deep Space Communication Complex today, showing the former Honeysuckle Creek dish in the foreground, with 70-metre DSS 43 behind it. The newer 34-metre antennae can be seen in the background. (Photo courtesy of NASA/CSIRO)

data and images from the Cassini spacecraft as it makes its final dive toward Saturn's cloud tops at the termination of its thirteen-year mission exploring the planet's complex system of moons and rings.

The last major NASA tracking station in Australia, CDSCC today is supporting more than 40 deep space missions from 25 nations. In 2017, a new control facility was added to the station, enabling it to take command of all three of NASA's deep space complexes for eight hours each day, in turn with the other DSN complexes. CDSCC's continuing presence in Australia enables the country to make an ongoing contribution to unlocking the secrets of the Solar System and the universe beyond, and all indications are that CDSCC and Australia will continue to do so for decades to come.

THE MANNED SPACE FLIGHT NETWORK: ENSURING ASTRONAUT SAFETY

In November 1958, NASA initiated its first human spaceflight program, Project Mercury. The small Mercury space capsule carried only one astronaut but it, together with the corresponding Soviet Vostok program, represented the first tentative steps in humanity's in-person exploration beyond the Earth.

At the time that the Mercury program was announced, the effect of space environment upon the human body was a completely unknown factor: it was not even certain that a person could endure the rigours of weightlessness for long periods or that the senses would function well enough for an astronaut to carry out mission tasks. To provide for the safety and well being of its astronauts in orbit, NASA established a dedicated tracking and communications network for the Mercury program that would keep astronauts in touch with mission control and the doctors who would be monitoring their medical condition. It would also provide for radio control of the spacecraft, to activate its re-entry systems, in the event that the astronaut became incapacitated.

Western Australia, already recognised as a suitable location for a tracking station to determine if an object launched from Cape Canaveral had achieved orbit, was also the best place for a control facility that could remotely command re-entry if an astronaut was unable to perform the task. Consequently, NASA established its only Mercury command station outside the United States at Muchea, 60km north of Perth.

Established in 1960, Muchea was tracking Station No. 8 in the Mercury network. Its equipment included radar and acquisition aid tracking systems, telemetry reception and air-to-ground voice communication facilities. During Mercury flights, an Australian doctor would serve as Flight Surgeon at the station, monitoring the medical telemetry from the spacecraft, which reported on the astronaut's health and physical condition, while one of the Mercury astronaut team acted as Capcom (Capsule Communicator), in direct voice contact with his colleague in orbit.

Since it was not required for the first two sub-orbital Mercury flights, Muchea's first operational mission was the tracking of Friendship-7, the spacecraft in which John Glenn made the first American orbit of the Earth, in February 1962 (following Soviet cosmonaut Yuri Gagarin's first orbit of the Earth in April 1961).

Because of the presence of Muchea, interest in Glenn's mission was particularly intense in Perth. The city decided to salute him as he passed over in orbit by turning on all the lights, a gesture that was also followed in the town of Rockingham, south of Perth. When Glenn asked Muchea about the source of the brilliant display that he could see, and was told that citizens in both towns had turned on their lights for him, he thanked them publicly from orbit, thus bringing Perth and Rockingham to world attention. In return for this friendly gesture, the Mayor of Perth was invited to attend ceremonies in New York honouring the first US Earth-orbiting astronaut. During the visit, Glenn quipped to the Mayor 'I hope I didn't run up your light bill!'

Following the end of the Mercury program, Muchea was closed, to be replaced by a new tracking station at Carnarvon, which would serve the Gemini and Apollo programs, but Australia's second Mercury facility, at Red Lake on the Woomera Range, would continue to operate until the end of the Gemini program.

Mercury Station No. 9 at Red Lake, 43km north of the Woomera Village, developed around a radar station built for the Blue Streak missile program.

The FPS 16 radar was seconded to NASA to support Mercury tracking and a small station, with voice communications and telemetry reception, was established close by. Red Lake supported Mercury and Gemini missions and also tracked some of the early automated Apollo tests before being retired as a NASA facility.

With NASA moving forward on the Apollo lunar landing project after 1961, Project Mercury was succeeded by the Gemini program. During 1965 and 1966, Gemini spacecraft carrying two astronauts tested the equipment and techniques, such as space rendezvous and spacewalking, necessary for accomplishing a mission to the Moon. To support the Gemini and subsequent Apollo flights, NASA greatly expanded the Mercury tracking network, re-naming it the Manned Space Flight Network (MSFN).

Because the orbits of Gemini spacecraft were different to those of Project Mercury, a new MSFN station was established at Carnarvon in Western Australia, to which equipment from Muchea was transferred. Officially opened in 1964, Carnarvon acted as both a MSFN station and a STADAN facility and, with the additional equipment installed for the Apollo missions, was the largest NASA tracking station outside the United States. Its equipment included a highly accurate FPQ 6 radar that was used to track crewed spacecraft and satellites and a Range and Range Rate system designed for tracking scientific satellites out to lunar distances. Early missions supported by this system included the Interplanetary Monitoring Platform, the Orbiting Geophysical Observatory series of satellites and the BIOS biological satellite program.

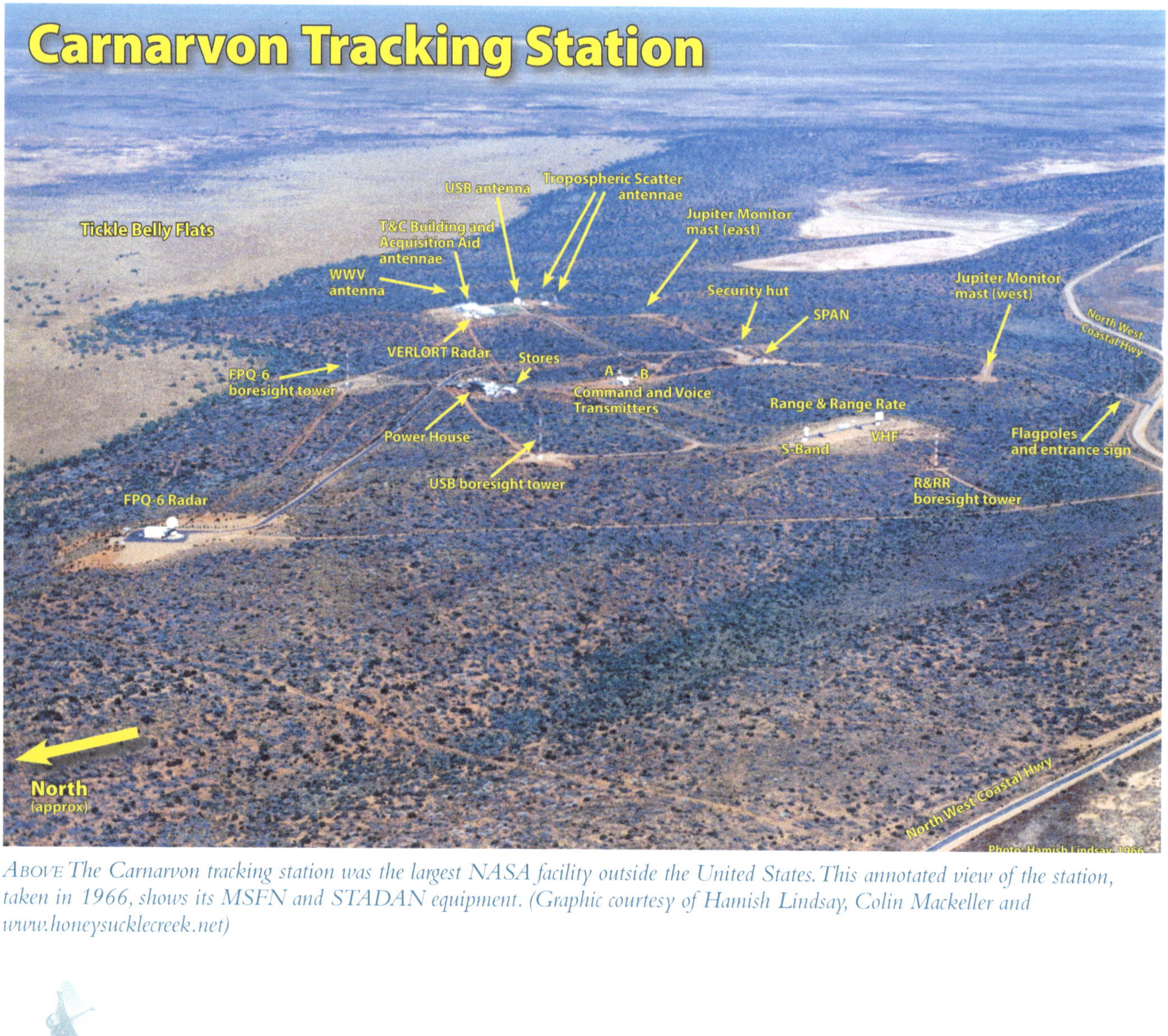

ABOVE The Carnarvon tracking station was the largest NASA facility outside the United States. This annotated view of the station, taken in 1966, shows its MSFN and STADAN equipment. (Graphic courtesy of Hamish Lindsay, Colin Mackellar and www.honeysucklecreek.net)

Gemini 3, in March 1965, was the first crewed mission supported by Carnarvon, which would go on to participate in every Gemini and Apollo mission, as well as the Skylab space station program in 1973-74. An astronaut Capcom was stationed at Carnarvon for Gemini 3, but there were no further astronauts assigned to this role for later missions, although RAAF doctors were based at the station as Flight Surgeons. During the Apollo program, Carnarvon's location gave it the critical role of determining whether a lunar mission would be given a go/no command to proceed out of Earth orbit and on its way to the Moon.

To ensure astronaut safety during the Apollo lunar missions, Carnarvon also hosted one of three stations in the world-wide Solar Particle Alert Network (SPAN), established to monitor the Sun during Apollo Moon missions. Certain types of solar activity, such as solar flares, spew out streams of charged particles that would be harmful to astronauts outside the protective screen of the Earth's ionosphere. Consequently, astronauts on Apollo lunar missions would be seriously at risk from radiation poisoning if a major solar event occurred during their spaceflight. In order to ensure that spaceflights were not launched during dangerous periods of solar activity, NASA established the SPAN observatories to monitor the Sun continually. Three stations — in Houston, Texas, the Canary Islands and Carnarvon — provided a 24-hour watch on the Sun.

Operating from sunrise to sunset every day, the Carnarvon observatory used two small optical telescopes to monitor solar activity: one telescope observed the Sun in the wavelength of hydrogen alpha, in order to detect solar flares; the other telescope was used to monitor sunspots. A small radio telescope was also used to 'listen' to the Sun at three different wavelengths, to detect signs of dangerous activity. The observatory also operated a riometer, an instrument that could detect the disturbances created in the upper atmosphere by atmospheric nuclear testing. This device could be used to warn Earth-orbiting crewed spacecraft of areas of dangerous radiation created by nuclear testing. The SPAN station was operated for NASA by the US Environmental Science Services Administration, which also operated a world-wide network of observatories to monitor the Sun for solar activity that would disrupt radio communications. Its two observers were employed by the Australian government.

Directly connected with the Carnarvon station's operations was Australia's first international satellite communications ground station, operated by OTC, which was established close by the NASA station. Given the often unreliable ground-based links connecting remote Carnarvon to the rest of Australia, this ground station (further discussed in Chapter 6) allowed the tracking station to directly communicate with the US via an INTELSAT satellite.

To provide additional tracking and communications support for spacecraft around the Moon during the Apollo program (1968-1972), Carnarvon was joined by an additional MSFN station at Honeysuckle Creek near Canberra. Because the Apollo lunar missions would involve two separate spacecraft in operation at the Moon (the Command Module, with one astronaut, remaining in lunar orbit while the other two crew members journeyed to the surface in the Lunar Module), the

BELOW Honeysuckle Creek MSFN station would bring the world the television signal of Neil Armstrong making his historic first step onto the Moon in July 1969. (Photo courtesy of Hamish Lindsay)

Australia and the First Moon Landing

When Apollo 11 carried the first astronauts to a landing on the lunar surface on 21 July 1969, Australian facilities played a critical role in providing live television coverage of the event. Carnarvon, Honeysuckle Creek and Tidbinbilla all played their part in 'landing a man on the Moon and returning him safely to the Earth', the goal that US President Kennedy had set NASA in 1961. The signals from Apollo 11 came through the Honeysuckle Creek and Tidbinbilla tracking stations, with assistance from the CSIRO's Parkes Radio Telescope, which was connected to the NASA stations by a microwave link, during the actual Moonwalk.

At the crucial moment of the first human footsteps on the Moon, Honeysuckle and Parkes were focused on the Lunar Module, Eagle, which was on the lunar surface, while Tidbinbilla was tracking the Command Module, Columbia, in orbit around the Moon. The Parkes Radio Telescope, with its large 64-metre dish, was originally enlisted by NASA to act as a back-up for the lunar television broadcasts, which were expected to be received by the large Goldstone antenna in California. However, changes to the mission flight plan elevated Parkes to the prime receiving station for the lunar surface broadcast.

Another last minute change meant that the Moon would only just be rising in Australia as Armstrong prepared to step out onto the lunar surface. Three tracking stations were receiving the signals simultaneously as Buzz Aldrin activated the television camera at 12:54:00 pm (AEST): Honeysuckle Creek, Goldstone and Parkes, which was able to receive signals on its less sensitive 'off-axis' detector, even though the Moon was too low in the sky there for them to be received in the main detector. The television from both Australian facilities was sent to Sydney via specially installed microwave links, where a NASA officer selected between the Parkes and Honeysuckle Creek signals to be relayed to Houston for inclusion in the international telecast. In Houston a controller then selected between the Goldstone and the previously selected Australian TV signals.

During the first nine minutes of the broadcast, NASA alternated between the Goldstone and Honeysuckle Creek signals, searching for the best quality images. They began with the Goldstone pictures, but the image was poor. Although the Honeysuckle Creek pictures were very grainy because of the low signal strength received by the smaller dish, they were the images transmitted to the world when Neil Armstrong took that 'one giant leap for Mankind'. Finally, eight minutes and 51 seconds into the broadcast Houston switched to the transmissions from Parkes, which was now relaying much higher quality signals, as the Moon had finally risen far enough to be in the field of view of the telescope's main detector. NASA then stayed with the Parkes television for the remainder of the 2.5-hour telecast.

At the time, Western Australia lacked a co-axial cable to bring television from the eastern states so that they, too, could watch the historic event in which their own tracking station was participating. Consequently, a complex operation was put in place to bring the television coverage from the United States back into Western Australia via the OTC Satellite Earth Station that supported the Carnarvon tracking station's operations. From Carnarvon, where a signal was specially broadcast into the town's Memorial Theatre so that locals could see the it, the vision was sent to Perth via a newly opened co-axial cable link, so that at least some parts of Western Australia could join the six hundred million people (about one-fifth of the world's population at the time) who watched Neil Armstrong's first steps on the Moon, relayed via Honeysuckle Creek.

Chapter 4 – Keeping on Track: Space Tracking From Australia

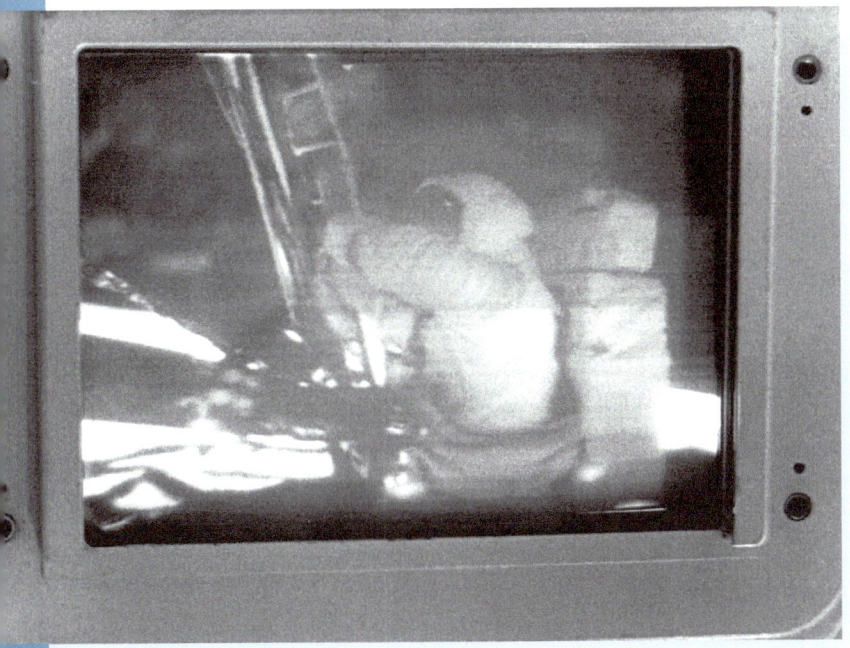

Above Neil Armstrong prepares to step onto the lunar surface, as seen on a monitor at the Honeysuckle Creek tracking station. The image is much clearer than those broadcast on television due to degradation caused by scan conversion and transmission methods. (Photo courtesy of Ed von Renouard and honeysucklecreek.net)

Honeysuckle MSFN station was sited deliberately close to the Tidbinbilla DSN station, so that the two facilities could work together while Apollo spacecraft were at the Moon, with Tidbinbilla also available to provide back-up in the event of any failure at Honeysuckle.

Honeysuckle Creek was equipped with a 26-metre antenna, the same as the antenna at Tidbinbilla. As NASA had adopted for the Apollo program the S-Band communications system developed by JPL for deep space missions, this meant that both stations used similar equipment. A dedicated second control room, the 'MSFN Wing', was added at Tidbinbilla, and the two stations were linked by microwave relay, so that they could work together. Honeysuckle Creek supported all the crewed Apollo missions, particularly earning its place in history by being the station through which the footage of Neil Armstrong stepping onto the lunar surface during the Apollo 11 mission was received and then relayed to the world.

Right Apollo 13 Command Module Pilot Jack Swigert visited Australia in 1972 to thank the tracking station crews who had helped to rescue the crippled mission. At Honeysuckle Creek MSFN station, Swigert (left) and US Ambassador Walter L Rice (with hat), meet with Station Director Don Gray. (Photo courtesy of Hamish Lindsay and www.honeysucklecreek.net)

All NASA's MSFN resources in Australia, together with the Parkes Radio Telescope, were stretched to the limit during the Apollo 13 space emergency, when a fuel cell explosion crippled the spacecraft, leaving the crew to rely on the Lunar Module as a 'lifeboat' to keep the astronaut alive until they could return to Earth. As Honeysuckle Creek had been tracking the spacecraft when the accident occurred, it was responsible for co-ordinating communication and tracking during the early stages of the rescue operation, with Carnarvon and the MFSN Wing at Tidbinbilla providing valuable support. A team from Tidbinbilla made an emergency dash to Parkes to re-install equipment that would enable the dish to track the stricken spacecraft.

A tracking beacon on the third stage of the Saturn V launch vehicle created interference, because it was operating on the same frequency as the Lunar Module, which under normal circumstances would not have been transmitting at the same time as the Saturn stage beacon. The 26-metre antenna at Honeysuckle Creek did not have a sufficiently narrow beamwidth to enable it to discriminate between the two, causing problems with tracking and signal reception. However, the greater sensitivity and narrower beamwidth of the Parkes antenna enabled it to isolate the Lunar Module signals, which were then passed on to Houston for analysis. The precise position of the spacecraft was determined and a new trajectory calculated to bring the crippled Apollo 13 and its crew home safely.

Like the STADAN station at Orroral Valley, Honeysuckle Creek received data from the ALSEP packages on the Moon. It also supported the Skylab space station missions. In 1974, with NASA's human spaceflight program on hold awaiting the development of the Space Shuttle, the Honeysuckle Creek station was transferred to the Deep Space Network, as DSS 44. During this period it supported planetary exploration missions such as the Viking Mars landings, the Voyager probes to the outer solar system, the Pioneer Venus missions and the joint West German/NASA Helios solar missions.

In 1981, the Honeysuckle Creek station was permanently closed, but its historic antenna was transferred to Tidbinbilla, where it would continue in operation as DSS 46, monitoring Earth-orbiting satellites, until 2009, when it was finally retired. Today, although the Honeysuckle Creek facility is gone, the site has been added to the Heritage Register of the ACT, along with the former Orroral Valley tracking station site, and informational signage describing their roles in the history of space exploration has been installed at both locations. The original Honeysuckle antenna at Tidbinbilla remains as a monument to Australia's contribution to the historic Apollo 11 lunar landing.

Other NASA Facilities in Australia

In addition to its major fixed space tracking facilities, NASA has established some temporary stations to support specific projects. The earliest of these was a 14-foot (four-metre) dish located near Darwin to support the Orbiting Geophysical Observatory series of satellites, six of which were launched between 1964-1969. Working in conjunction with the Carnarvon station, the Darwin station was in operation from 1965-1969 before being moved to the MSFN station in Hawaii.

In 1966, a temporary STADAN station with a 12-metre dish was established at Cooby Creek dam, about 22km north of Toowoomba in Queensland, in conjunction with the Applications Technology Satellite (ATS) series. This program was used to investigate new techniques for satellite communications, meteorology and different kinds of spacecraft stabilisation systems. The first weather satellite image showing the whole of Australia was obtained from ATS 2 in April 1967.

Through ATS 1, Cooby Creek brought the first live satellite television broadcasts into Australia. These included the special 'Australia Day' events at Expo 67 in Montreal, Canada, in early June 1967, followed on 26 June by 'Our World', the first global television broadcast. Highlights of the broadcast from Expo 67 included performances by The Seekers, Normie Rowe, Rolf Harris and Bobby Limb. For the 'Our World' broadcast, which commenced at 4.50am EAST, Australia contributed footage of the first tram of the day leaving a depot in Melbourne, a tour of the CSIRO's 'Phytotron' plant growth laboratory in Canberra and a visit to the Parkes Radio Telescope, where viewers saw, and heard, an observation of the most distant object then known – Quasar 0237-23. In December 1967, Cooby Creek relayed to the world live coverage of the memorial service for Prime Minister Harold Holt, who had disappeared while swimming at Cheviot Beach on Port Phillip Bay on Sunday 17 December 1967. A few weeks later, the 1967 Davis Cup tennis final from Brisbane was also broadcast to the world.

Although NASA completed its ATS experiments in Australia in 1969, the Cooby Creek station remained in Australia for an additional year, on loan so that the PMG and the Department of Civil Aviation (DCA) could conduct experiments. The PMG was interested in investigating the use of satellites for telephone and distance education services to remote communities and pastoral stations, experience which would later feed into the development of the Homestead and Community Broadcast Satellite Service (HACBSS) for the Aussat system. The DCA experiments involved using satellites for aircraft communications and ranging, the latter using a special transponder installed on test aircraft. When these experiments were concluded, the mobile station was finally removed and redeployed to Spain for another NASA program.

Since 1979, NASA has also maintained a laser ranging station near the town of Dongara, Western Australia, about 350km north of Perth. Mobile Laser Station (MOBLAS) 5, located on a property called Yarragadee (which has also been applied as a name to the laser station), was established to support NASA's Earth Dynamics program, which uses lasers, reflected from the Moon and specially equipped laser-reflecting satellites to make very accurate determinations of distance. This can assist in a wide range of studies such as polar motion and Earth rotation, the orbit and interior structure of the Moon, and studies of the tides and gravity.

The station plays an important part in Australian and international geodesy studies and works in conjunction with Australian national geodetic facilities. Originally managed and operated for NASA by the then-Division of National Mapping, MOBLAS 5 is now operated by the Division's successor agency, Geoscience Australia and works in conjunction with its laser ranging facilities. The site is also the location of the University of Tasmania's Very Long Baseline Interferometry (VLBI) radio telescope (further information on laser ranging and VLBI can be found in Chapter 7), and hosts time reference equipment and a transmitter for the French-developed DORIS (Doppler Orbitography and Radiopositioning Integrated by Satellite) system. DORIS is a civil precise orbit determination and positioning system based on the principle of the Doppler effect. It employs a network of transmitting terrestrial beacons and onboard instruments carried by satellites.

ESTRACK: SUPPORTING ESA'S SPACE OPERATIONS

Like NASA, the European Space Agency needed a global network of tracking stations that would form its equivalent to the STADAN network. Known as Estrack, the ESA network was first established in 1975, operated from the European Space Operations Centre (ESOC) in Darmstadt, Germany. Estrack was developed to provide tracking through all phases of a space mission, from 'LEOP' – the critical Launch and Early Orbit Phase – through to routine operations, special manoeuvres or flybys and eventually the deorbiting and safe disposal of a satellite to reduce the problems of space debris.

RIGHT Image from the ESA comet probe Giotto, showing the nucleus of Comet Halley. The Australian Estrack station, with the support of the Parkes Radio Telescope, controlled Giotto's encounter with Comet Halley in 1986. (Photo courtesy of ESA)

As ELDO's successor, the European Space Agency planned to launch its satellites from the French launch facility in Kourou, French Guiana, to which ELDO had switched its operations from Woomera after 1970. In the same way that spacecraft launched from Cape Canaveral could be monitored by a tracking station situated in Western Australia, so too could any launch from Kourou. Consequently, in 1979 ESA established a tracking station for launch and Earth orbit support at Carnarvon, locating it alongside the OTC Satellite Earth Station, which managed the station on ESA's behalf.

The initial facility consisted of only two VHF antennae, but this was soon joined by a 15-metre S-band dish, installed to support ESA's first deep space probe, the Giotto mission to Comet Halley in 1986. The Carnarvon station was the prime command facility for Giotto's encounter with the comet's tail, which was also supported by the Parkes Radio Telescope, lending its capabilities to ESA as it did to NASA. Other missions supported by Estrack and OTC Carnarvon included the launch and orbital positioning of the European geostationary weather satellite Meteosat 2; control of the Marecs communication satellite;

providing launch support for the launch of India's Rohini satellite in 1980 (the first Indian satellite to be launched on an Indian-made rocket) and monitoring the launch of Japan's MOS 1 marine observation satellite in 1987.

When the OTC station at Carnarvon closed, in 1987, the Estrack station moved to the OTC's new Perth International Telecommunications Centre at Gnangara, where it operated until the end of 2015, being then retired due to urban expansion and the increased risk of radio interference. Instrumental in rescuing the Hipparchos astronomy satellite after it failed to achieve its proper orbit in 1989, the Perth Estrack station supported Ariane, Vega and Soyuz launches from Kourou, and some of ESA's highest profile Earth observation, science and navigation missions, such as XMM-Newton, SMART 1, the Sentinel remote sensing satellites and the four identical Cluster-2 satellites studying the impact of the Sun's activity on the Earth's space environment. It also assisted missions from ESA partner agencies including the French space agency and Eumetsat, the European meteorological satellite consortium.

Significant operations in its final years before closure included: monitoring the double launch of ESA's Herschel and Planck space telescopes in 2009; tracking both the launcher and satellite simultaneously for the Gaia space observatory launch in 2013; and bringing the LISA Pathfinder gravitational wave detector spacecraft through six critical orbit-raising manoeuvres to reach its final orbit 1.5 million kilometres from Earth in 2015.

While the retired Gnagara antenna was moved to the Azores, a new 4.5-metre antenna was installed at ESA's deep space tracking station at New Norcia in Western Australia, to take over satellite acquisition and tracking duties. Some routine satellite tracking and control tasks for European satellites have been contracted out to commercial service providers, including SSC Australia, which operates the Western Australia Space Centre near Dongara.

In 1998, ESA decided to expand Estrack's capabilities by establishing its own deep space network, to handle the increasing number of planetary missions it had planned for the beginning of the Twenty-First Century. Like NASA's DSN, it consists of three stations located 120° apart in longitude. The first ESA Deep Space Antenna, DSA 1, was opened near New Norcia, a tiny Western Australia community about 150km north of Perth. Its sister stations are located at Cebreros (Spain) and Malargüe (Argentina). All three stations have a 35-metre antenna as their main instrument, operating in S and X-band, although there are plans to upgrade them to handle Ka-band, which has been designated as the new frequency for deep

Above ESA's Deep Space Antenna-1 tracking facility established near New Norcia in Western Australia. (Photo courtesy of ESA)

space operations. DSA 1 is also equipped with Delta DOR (Delta Differential One-Way Ranging), a new technology enabling highly precise spacecraft location and tracking and hosts equipment allowing scientists to analyse received signals to perform radio science experiments.

DSA 1 was completed in 2002, with operational trials conducted with NASA's Stardust comet mission, before being officially opened in March 2003. The station is remotely controlled from ESOC and operated by Inmarsat Solutions B.V., who provide on-site maintenance and additional support staff when required. Among the missions already supported by DSA 1 are Mars Express, Rosetta, Gaia and Cluster 2: in 2018 it will track BepiColombo, a joint ESA-Japan mission to Mercury.

With the installation at New Norcia in 2015 of the 4.5-metre antenna used for acquiring the first signals from newly launched satellites, the two dishes have now been connected, so that the 4.5-metre dish, with its wider field of view, can 'slave' point the larger antenna for orbital satellite tracking. When used in this way, the large antenna is designated NNO 1, with the smaller antenna designated NNO 2.

Just as NASA has used its tracking networks to assist the space missions of other nations, ESA shares the Estrack network with other space agencies under reciprocal agreements, which allow it to use their tracking resources in return. These include networks and stations operated by France, Italy, Germany, Japan and NASA. CDSCC has supported ESA missions including Huygens, Venus Express, Mars Express and Rosetta, while Estrack stations in Australia have assisted missions operated by China and Russia, as well as tracking the descent of NASA rovers to the surface of Mars.

Since 1957, Australia's involvement in space tracking programs has enabled it to develop considerable expertise in space tracking technology and to participate not only in the space programs of the United States, but also those of Europe and other nations, in a manner that would not otherwise have been possible.

NASA AND THE PARKES RADIO TELESCOPE

One of the world's major radio telescope facilities, responsible for many significant discoveries in radio astronomy, the 64-metre diameter Parkes Radio Telescope, was designed and operated by the Commonwealth Scientific and Industrial Research Organisation (CSIRO). Completed in 1961, it is located in the central western New South Wales town of Parkes, 365km west of Sydney. Although designed as a radio astronomy instrument and not a tracking facility, its capabilities have made it well suited for deep space tracking work and its large, fully steerable, dish antenna design was adopted and adapted by NASA for the 64-metre dishes of the Deep Space Network.

Tracking the first Pioneer probes into interplanetary space in the late 1950s made it evident to NASA that monitoring spacecraft at lunar and planetary distances required the largest possible antenna, leading JPL to consider the construction of much larger tracking dishes — in the 50–80-metre range — to augment its first 26-metre DSN antennae. The tracking characteristics that NASA determined it would need for these large antennae were closely approximated by the unique and innovative design of the Parkes Radio Telescope. The Parkes telescope's master equatorial precision pointing system (conceived by the noted British consultant engineer, Barnes Wallis, of World War II 'dam busters' fame) provided a level of pointing accuracy that was precisely what the DSN required in order to maintain contact with distant spacecraft.

Consequently, even while the Parkes telescope was under construction, NASA approached the CSIRO about the possibility of using it for short-term tracking duties where an extremely strong, reliable

NASA and the Parkes Radio Telescope

signal was desirable. In 1961 and again in 1966, JPL actually requested the formal inclusion of Parkes in the DSN, but this was reluctantly declined by the CSIRO on the basis that routine space tracking duties did not warrant the displacement of the astronomical research being conducted by the radio telescope.

As part of its plan to base its new 64-metre antennae on a modified version of the Parkes design, NASA awarded a research grant to CSIRO in 1962, to determine and report on the detailed characteristics of the newly commissioned Parkes telescope. During this period, an excellent working relationship was established between the CSIRO and JPL, which proved to be of critical worth in Parkes' support of future space missions. Funds from the NASA grant were also used to cover the radio telescope's support for both Mariner 2 and 4, which NASA felt was vital for their encounters with Venus and Mars, augmenting the existing capabilities of the DSN.

Although Parkes experienced initial difficulties in finding and locking on to the signal from Mariner 2 as it approached Venus in December 1962, the staff there were able to overcome these problems and successfully tracked the spacecraft from late December 1962 until the signals ceased in January 1963. During the Mariner 4 Mars fly-by in July 1965, Parkes tracked the space probe from 8 July to 27 August, regularly recording its telemetry. During Mariner 4's approach to Mars, Parkes participated, in conjunction with Tidbinbilla, in important observations as the spacecraft emerged from behind the planet, aimed at probing the atmosphere and ionosphere of Mars. Parkes also received the delayed transmission data for the 22 images of the Martian surface captured by Mariner, providing higher quality images of the Martian surface than those received at the smaller DSN stations.

As already outlined, Parkes played a significant role in the Apollo 11 mission, with the radio telescope staff resolutely operating the antenna well outside its safety parameters during the lunar television broadcast, due to a violent windstorm, which struck the area at the time, in order to bring the historic vision to the world. Parkes' role in the first Moon landing became the basis for the popular, if not quite historically accurate, Australian comedy film, The Dish.

Following the success of Apollo 11, Parkes was again contracted to provide backup support for the Apollo 12 missions. In recognition of the radio telescope's contribution to these missions, NASA made another grant to the CSIRO, which was used to re-surface the antenna, allowing the telescope to operate more efficiently at higher frequencies. While not intended to participate in the Apollo 13 mission, Parkes was called in when the emergency occurred, its greater

LEFT Completed in 1961, CSIRO's Parkes Radio Telescope became the prototype for the 64-metre antennae of the Deep Space Network. (Image Copyright: CSIRO/CASS)

sensitivity allowing the extremely weak voice signals from the damaged Command Module to be received. As related above, it was also able to disentangle the Lunar Module's signals from those of the Saturn V's third stage beacon. Parkes was not required for Apollo 14, although for Apollo 15 it again played a vital role during critical phases of the mission and during Apollo 16 and 17 it tracked the spacecraft for very short periods to receive fast dumps of recorded data.

In the 1980s, Parkes once again became involved in planetary exploration, assisting ESA to receive signals from the Giotto spacecraft at Comet Halley in 1986, and returning to DSN support in the same year, to receive the very weak signals from the Voyager 2 spacecraft at Uranus. To facilitate future deep space co-operation between Tidbinbilla and Parkes, as well as enabling the two to undertake astronomical research using radio interferometry, a permanent microwave link was installed between them at this time. When undertaking joint interferometry research, Parkes and Tidbinbilla formed what was initially the world's largest real-time interferometer radio telescope.

Parkes assisted the DSN once again when Voyager 2 visited Neptune in 1989, and in 1996 was called upon to support the Galileo mission to Jupiter for a year. The failure of the Galileo spacecraft's main antenna to properly unfurl meant that data could only be transmitted via a back-up telemetry antenna and the signal reaching the Earth was extremely weak. To even extract the signal from background static, several DSN tracking antennae were linked and their signals combined: during the most favourable periods for signal reception, Parkes was linked to the 70-metre antennae at Goldstone and Tidbinbilla and to two of Tidbinbilla's 34-metre antennas, giving five antennae pointing at Galileo at the same time. This technique enabled 70 percent of the planned science data to be received and brought the world Galileo's amazing images of Jupiter and its moons.

To assist with a particularly busy period of deep space activity in late 2003-early 2004, Parkes was again incorporated into the DSN, being upgraded by NASA with new receivers and equipment capable of handling the transmissions from more advanced robotic explorers. During this period, it supported the exploration of Saturn's moon Titan by the ESA Huygens probe, which had hitched a ride to Saturn with NASA's Cassini mission. More recently, Parkes participated in experiments receiving signals from the Mars Exploration Rover Opportunity, as a test in advance of the Mars Science Laboratory Curiosity rover's landing on Mars in August 2012. During the landing itself, it joined Tidbinbilla and New Norcia in listening for signals that would indicate the safe descent and arrival of the rover on the Martian surface, after its risky landing procedure.

Today, Parkes remains ready to 'lend an ear' to the Deep Space Network or the planetary exploration missions of other agencies, if its capabilities are required.

BELOW *The Parkes Radio Telescope points toward the Moon during a rehearsal for the Apollo 11 mission a few weeks later. (Image Copyright: CSIRO/CASS)*

Above The launch vehicle used for the Project SPARTA re-entry experiments. Its Redstone missile first stage is still in its US Army livery. (Photo courtesy of Defence Science and Technology Group)

Right DST Group's Buccaneer cubesat project will contribute to Australia's defence by helping to improve the JORN radar surveillance network. (Photo courtesy of Defence Science and Technology Group)

CHAPTER 5

The High Ground: Using Space for Defence and National Security

For at least 2500 years, military strategists have recognised the importance of the 'high ground' – readily defended elevated terrain from which enemy movements can be observed. Even before the first satellite was launched, space was recognised as the ultimate high ground, since a satellite in orbit can observe broad swaths of the Earth's surface, gathering vital national security, strategic and tactical intelligence, while being virtually impervious to attack from the ground. Long-range missiles such as IRBMs and ICBMs also traverse space in sub-orbital trajectories as they travel from launch to their targets, releasing their warheads to re-enter the atmosphere at hypersonic velocities.

BACKGROUND IMAGE THIS PAGE *Under the radome – a view of the largest satellite antenna at the former Joint Defence Facility Nurrungar, one of three major US military tracking stations in Australia. (Author's collection)*

During the latter half of the Twentieth Century, impelled by the Cold War, all the major powers increasingly employed the 'high ground' view from orbit for defence and national security, establishing satellite networks for early warning, surveillance, communications, intelligence gathering and navigation. These space-based assets are now fundamental components of modern warfighting, security and defence, alongside the offensive and defensive capabilities of long-range missiles.

Australia's alliance partnerships with the United Kingdom and United States drew the country early into military space research, also enabling it to utilise their defence and intelligence satellite networks for its own defence and security. In exchange, Australia has allowed the United States to take advantage of its geographical location for its own satellite ground station needs – an arrangement that has not been without controversy.

Black Knight: Defence Research 'Sounding Rocket'

As Woomera's origins lay in weapons testing, it is not surprising that Australia's participation in military space research commenced with Britain's decision to develop the Blue Streak missile, which required a re-entry vehicle that would protect its nuclear payload from being incinerated by the fierce heat generated as it fell back into the atmosphere. To carry out this research, a smaller missile called Black Knight was developed to test the aerodynamic behaviour of re-entry nosecones of different profiles and materials.

Standing just over 10 metres tall, the Black Knight was the first space launcher designed in the United Kingdom. Although often referred to as a 'sounding rocket', since its research projects required it to fly ballistic trajectories like those of sounding rockets, the Black Knight was technically capable of launching a small satellite into orbit. It was, in fact, proposed as a component of several British satellite launcher designs, including the unsuccessful Commonwealth launcher proposal that preceded the formation of ELDO.

Like Blue Streak, the Black Knight's development commenced in 1955. Designed to conduct high-altitude, high-velocity re-entry flight tests, the Black Knight was first launched at Woomera in September 1958. By 1960, it had won acceptance as the Western world's most reliable rocket: there would never be an inflight failure in its 22 firings. Despite the cancellation of Blue Streak as a missile in 1960, Black Knight remained in service until 1965 for a series of 're-entry physics' experimental programs.

Early Black Knight flights revealed an unexpected phenomenon: the ionised gas tail of test heads re-entering the atmosphere returned strong radar echoes. Since this phenomenon had both offensive and defensive military implications and

LEFT *The single-stage Black Knight rocket, used for Project Gaslight. Two rockets can be seen here, ready for launch, with test heads of different shapes. (Photo courtesy of Defence Science and Technology Group)*

OPPOSITE PAGE *A replica of the Project Dazzle two-stage Black Knight on display in the Woomera Village Rocket and Missile Park. (Author's collection)*

Chapter 5 – The High Ground: Using Space for Defence and National Security

was thought to be the key to developing an anti-ballistic missile (ABM) system, further research was considered a priority. With the cancellation of the Blue Streak missile, Black Knight was devoted entirely to studying this phenomenon and related re-entry physics research, with the exception of one flight as a test vehicle for the ELDO Europa launcher.

Because the United States had observed this same radar echo phenomenon during its own missile research, its Advanced Research Projects Agency (ARPA) agreed in 1959 to participate in a tripartite collaboration with Australia and Britain to investigate it further. This resulted in two hypersonic re-entry physics research programs using Black Knight rockets at Woomera. The first, Projects Gaslight (1960-62), employed the original single stage Black Knight. However, for the second research campaign, Project Dazzle (1964-65), a two-stage version of Black Knight was developed, using a British Cuckoo rocket motor as an upper stage.

The results obtained from these tests contributed not only to missile development, but also to the development of the Gemini and Apollo spacecraft. Though the Black Knight vehicle was retired in 1965, its technology lived on in Britain's successful, but short-lived, Black Arrow satellite launcher discussed in Chapter 3.

Project SPARTA: Black Knight's Successor

As the requirements of Project Dazzle had stretched the capabilities of the Black Knight to the limit, ARPA proposed that further research should be undertaken with a launch vehicle provided by the United States. This follow-on program was designated Project SPARTA (Special Anti-missile Research Tests, Australia) and established in March 1966 under an agreement between the United States, Australia and the United Kingdom.

The Space Research Euphemism

Because of the highly classified nature of much of the weapons testing undertaken at Woomera – and, with it, the fear of Communist espionage or sabotage – from the earliest days, public information released about the research and tests conducted there was couched in ambiguous and euphemistic terms, describing many weapons development projects simply as 'rocket research'.

When space activities were initiated at Woomera under the IGY, the same approach was adopted to deal with the strong media pressure for access to the Range due to the intense public interest in space. 'Space research' became a convenient civilian-sounding euphemism behind which to conceal the military nature of many of the research programs that commenced in the late 1950s, associated with the development of the Blue Streak missile. From 1958, a significant number of the space launches at Woomera were of 'sounding rockets' involved directly, or indirectly, with space-related defence research.

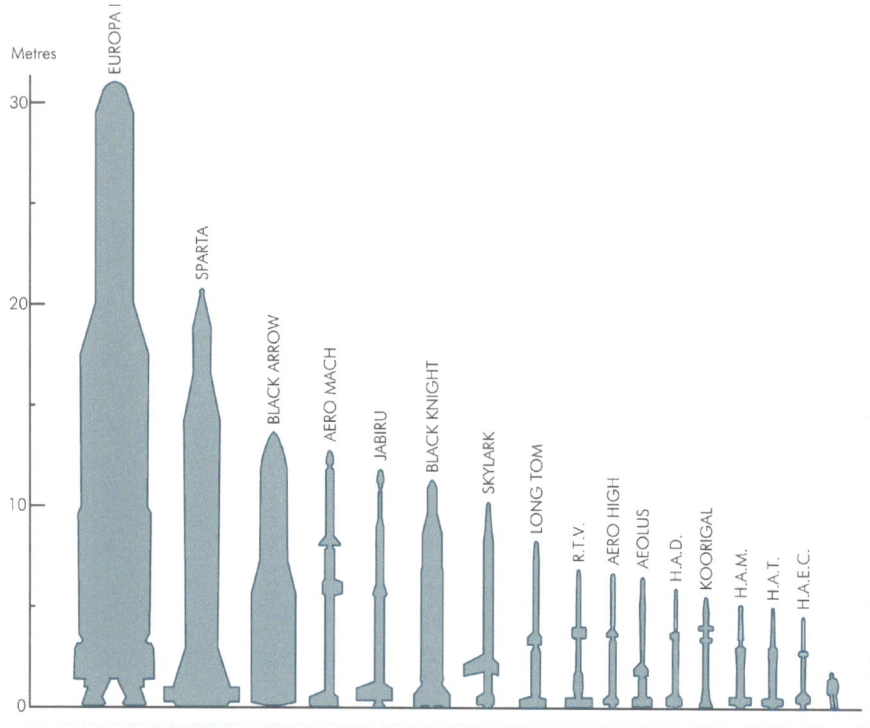

LEFT 'Research Rockets at Woomera', based on an original WRE diagram from the 1960s. Of those illustrated, SPARTA, Aero Mach, Jabiru, Black Knight, RTV, Koorigal, HAEC (and possibly also HAM) were actually involved in defence-related projects.

The first stage of the SPARTA vehicle was the Redstone missile, which could trace its development lineage directly back to the German V 2. Despite being operationally obsolete by 1966, the Redstone had already been successfully adapted as a satellite and human-rated launcher. It was well-suited to the proposed SPARTA research program, in which test heads, after being launched into space, would be driven back into the atmosphere at extremely high re-entry velocities by the vehicle's upper stages. Ten Redstone missiles were provided by the US Army, which was forced to borrow back the support equipment from the Smithsonian Institution, to which it had been consigned after the Redstone was retired.

For Project SPARTA, two small solid-fuel upper stages were added to the Redstone: an Antares 2 rocket formed the second stage, while the third stage was a US-made BE-3, modified by the WRE for the project. A series of nine launches was planned, but a tenth vehicle was brought to Australia as a spare, in the event of launch failures. An American launch team would prepare each rocket for flight at Woomera, where the project was assigned to Launch Area 8, a site that would also be used for the launch of WRESAT and later for Aerobee sounding rockets for US and Australian upper atmosphere research. The earlier Black Knight firings had taken place at Launch Area 5, which was also used by the Black Arrow satellite launcher.

Although the first SPARTA launch in November 1966 was a partial failure, the second flight a few weeks later was considered a success – so much so that it was felt very unlikely that the spare vehicle would be needed. Since ARPA had no interest in returning the obsolete missile back to the United States, it was open to the suggestion that this spare be used for the Australian satellite project that became WRESAT, as outlined in Chapter 3.

The last large rocket program for hypersonic re-entry tests during the Joint Project, data from SPARTA contributed to both missile programs and the development of the Apollo spacecraft heatshield, just as the earlier Projects Gaslight and Dazzle contributed to spacecraft development.

HYPERSONIC RESEARCH ROCKETS: JABIRU AND FALSTAFF

Alongside the major missile re-entry vehicle tests and ionisation physics experiments carried out with the Black Knight and under Project SPARTA, a broader program of hypersonic research commenced at Woomera in the late 1950s. The term 'hypersonic' refers to velocities greater than Mach 5 (five times the speed of sound) and an understanding of hypersonic aerodynamics was of particular importance not only to the design of missile warheads, but also to the development of early aerospace vehicles, such as NASA's 'lifting bodies', and high performance fighter aircraft.

Jabiru

Because hypersonic conditions were beyond the simulation capabilities of the wind tunnels of the day, hypersonic research rockets were specifically designed to allow the study of warhead and scaled down vehicle mock-ups during atmospheric re-entry. In 1958, the UK's Royal Aircraft Establishment (RAE) and the WRE began collaboration on the design and development of a hypersonic research rocket known as Jabiru (also initially called Jaguar in the UK), establishing the Hypersonic Research Vehicle (HRV) Joint Project, which would continue until 1970.

The initial Jabiru vehicle was composed of a British first and second stage, while the third stage was an Australian-developed rocket known as Lobster, a descendant of the original Australian Marloo motor discussed in Chapter 2. However, performance issues with the Lobster meant that it was replaced by another British rocket motor for the Jabiru Mk. 2 (also known as Aero Mach), introduced in 1964. The research program for the ten flights of the Jabiru Mk.1 was focused on measurements of heat transfer and pressure distribution in re-entry vehicle designs of various shapes, while the follow on Aero Mach program encompassed materials research and aircraft wing design in addition to missile nosecone testing.

Within the Aero Mach program, three single-stage vehicles were allocated to launching into free-flight an aircraft model named Oberon. Developed by the RAE, the Oberon was intended to simulate a highly swept-wing Mach 4.5 aircraft design. The WRE designed and built the model support and separation system, the aerial systems for the rocket stage and the first two 200 lb (90kg) simplified models to check out the complete vehicle.

Despite the formal termination of the HRV Joint Project in 1970, the RAE and WRE continued to collaborate on hypersonic research, with additional Jabiru Mk. 2 flights in 1971, and the introduction of a new two-stage vehicle, Jabiru Mk. 3, in 1974. By the time the last Jabiru rocket was fired in 1974, British and Australian interest in hypersonic research had waned and both wind tunnels and computer capabilities had advanced to the stage where some research could be undertaken without the need for expensive rocket launches.

Above The Jabiru Mk.1 Anglo-Australian hypersonic research rocket. Its third stage was the Australian-developed and manufactured Lobster rocket motor. (Photo courtesy of Defence Science and Technology Group)

Left Hypersonic research conducted at Woomera contributed to the design of missiles, spacecraft and experimental aerospace vehicles like NASA's M2-F2 lifting body. (Photo courtesy of NASA)

Across its lifespan of some 15 years, the data gathered from Jabiru firings contributed to the larger body of data on hypersonic aerodynamics and ablation effects that was amassed at Woomera from British, Australian and American research programs. The results of this research were incorporated into the design of missiles, aircraft and spacecraft.

Falstaff

The post-1970 Jabiru trials are believed to have been connected to the development of the British Polaris missile upgrade program designated Chevaline, which was considered so secret that it was not revealed until 1980. After abandoning its own IRBM development program with the cancellation of Blue Streak, the UK had purchased American Polaris submarine-launched missiles for its nuclear deterrent, but preferred to develop its own warhead system. The Chevaline program was intended to upgrade the re-entry vehicle's defence system for its multiple warheads with penetration aids and decoys that would enable at least some of the nuclear payloads to survive interception by an ABM system.

BELOW ELDO Launchpad 6A at Woomera was scrapped and demolished during the 1970s. Other Range facilities were allowed to deteriorate during the 1980s and 90s. (Author's collection)

Additional Chevaline research, requiring a more powerful launch vehicle, was undertaken with the Falstaff launcher, based upon the first stage of a hypersonic research vehicle called Hyperion, which had been cancelled in 1969. Known as Stonechat, this first stage was the largest solid-fuel rocket motor ever flown by the UK. The first Falstaff launch occurred in 1969, to test the Stonechat stage, but the operational program did not commence until 1975, following a period of political uncertainty about the Chevaline project in Britain. Ultimately, seven test flights were made between 1975 and 1979 using the Falstaff vehicle, which combined the Stonechat booster with protective payload fairings based on those used for the Black Arrow. The last Falstaff flight, in April 1979, marked the end of hypersonic research at Woomera until the Twenty-First Century, as well as the termination of the Anglo-Australian Joint Project, which came to a close in 1980.

THE AFTERMATH OF THE JOINT PROJECT

With the cessation of the Joint Project looming, and having no intention to pursue missile development or space projects on its own, the Australian Government began to refocus its defence science efforts from the mid-1970s, moving toward providing a broader research and development capability that would serve Australia's defence needs as a whole.

To facilitate this new policy, the ADSS was restructured into the Defence Science and Technology Organisation (DSTO) in 1974, becoming integrated with the in-house research and development units of the Armed Services and the Science Branch of the Department of Defence. All Defence research and development activities were transferred to DSTO, which, in 1975, came under the control of the Department of Defence when the former Department of Supply was dissolved.

In 1978, the Weapons Research Establishment, which remained separate from DSTO, was itself restructured, to become the Defence Research Centre Salisbury (DRCS). Although it retained the management of the Woomera Range, the oversight of the NASA tracking stations in Australia had been removed from the WRE as early as 1972, when it was transferred to the Department of Supply's Canberra headquarters. With the Department of Supply itself dissolved in 1975, the NASA oversight function became the responsibility of a new Space Projects Branch at Department of Science, before being transferred again to the newly-created Australian Space Office in the late 1980s (see Chapter 9 for further details). In July 1987, the Defence Research Centre became part of the Defence Science and Technology Organisation, which then assumed the management of the Range, which had been renamed Woomera Test Facility (WTF) in 1980 and would later be called the Woomera Test Range (WTR).

Woomera Comes Full Circle

With the end of Joint Project, Woomera's use by the Australian Defence Force (ADF) rapidly declined: by the latter half of the 1980s, the Range was being used, on average, only for about ten weeks a year for such tasks as Global Positioning System tests, explosive tests, vehicle and missile tests and an occasional sounding rocket launch. Much of the Range's plant and equipment was sold off, scrapped, or abandoned to deteriorate in situ. Woomera Prohibited Area was gradually scaled back from its peak area of 270 000sqkm (100 000 sqmi) to its current total area of 122 188sqkm (47177sqmi), although even at this reduced size it is still the largest land-based weapons test range in the Western world. In 1982, the Woomera Village, which had been a restricted area since it was first constructed, was opened to the public in the hope that tourism would bring much-needed revenue into the town as the population dwindled.

By 1989, the high cost of maintaining Woomera led the Hawke Government, which was unwilling to completely abandon Range due to the enormous capital investment it represented, to announce a proposal for its commercialisation. The Government sought to find a commercial developer to meet the cost of maintaining and developing the Range to service new markets, both military and civilian, including the resources sector. Despite a suggestion that the WTF might be redeveloped as a major facility offering air combat manoeuvring, electronic warfare and advanced bombing test facilities, missile testing and rocket launching on a commercial basis to military forces in the Asia-Pacific region, the attempt at commercialisation did not succeed and management of the Range transferred to the RAAF Aircraft Research and Development Unit (ARDU) in 1991.

Although there were several proposals for the revitalisation of Woomera as a civil space launch facility for light launch vehicles and re-usable spaceplanes across the 1990s (which will be further discussed in Chapter 8), none came to fruition. Japan, though, did conduct thirteen tests of its ALFLEX (Automatic Landing Flight Experiment) technology demonstrator vehicle at Woomera in 1996. This test vehicle, part of Japan's ultimately cancelled HOPE spaceplane development program, was released from a helicopter flying at high altitude to approach and land automatically on a runway. Japan would return to Woomera in 2002 and 2005, in order to use the sounding rocket facilities for test flights of a scale model of its NEXST 1 supersonic transport aircraft, investigating ways to improve fuel efficiency.

Largely idle through the 1990s, the Range and its facilities continued to deteriorate, with the closure of the nearby Joint Defence Facility Nurrungar (to be further discussed below) coming as a particular blow to the Woomera township, where its personnel had been based. Seeking to bolster the economy of the Village, the Howard Government established the Woomera Immigration Reception and Processing Centre, a highly controversial detention centre for those the Howard Government considered illegal immigrants. Opened in 1999, the trouble-plagued facility was closed in 2003.

However, in 1999, ARDU recognised that changes in the global strategic environment and new developments in weapons technology meant that Woomera had renewed potential as a weapons testing and evaluation range, being both electromagnetically quiet, and thus suitable for electronic warfare research and testing, and one of the few sites in the world where over-the-horizon weapons testing was feasible, due to its downrange length of some 630km.

National and international promotion of these characteristics brought a growing number of military, defence industry and also civil users to the Range, so that by 2009, multiple test areas were in operation daily and bookings for Range access had been made for more than ten years ahead. This increasing use and evidence of long-term interest in the Woomera facility prompted the Australian Government to embark, in 2009, on a major refurbishment and upgrade of Woomera's tracking systems and other infrastructure.

After decades of underuse, Woomera has now returned to its original role as a weapons testing facility, under control of the RAAF. Renamed the Woomera Range Complex (WRC) in 2013, the Range is now used for Australian Defence Force trials, and for the testing of weapons systems and Unmanned Aerial Vehicles (drone aircraft) by foreign militaries and private companies. Its facilities include: a fully instrumented air weapons range; demolition ranges; live firing ranges and target areas for both aircraft and artillery, and sounding rocket launch capabilities. In 2016, a new program of upgrades commenced, intending to make Woomera the world's most advanced test range.

Although this revitalisation of Woomera does not currently include any major space projects, the Range's potential to host light satellite launch vehicles, airborne satellite launchers, sub-orbital vehicles for space tourism or commercial transport, sounding rockets and hypersonic research remains, and could still be realised at some time in the future.

Defence Space Capabilities

Since the 1980s, alongside the growth of civil satellite applications that will be discussed in Chapter 6, governments and defence forces across the globe have come to rely increasingly on space-based assets, in the form of dedicated military satellite networks and data from civil and commercial satellite services, to enhance their warfighting capabilities and support national security and defence. Australia is no exception to this trend and its Defence and Intelligence agencies make extensive and sophisticated use of space for communications, navigation, timing and positioning services, early warning, intelligence, surveillance and reconnaissance, and meteorological, environmental and geospatial information to support its defence activities and formulate government policy on national security issues.

Above Logo of the Woomera Test Range, the major component of the Woomera Range Complex under RAAF management. The Range's current motto 'Sharpen the spear' refers to the role that Woomera plays in enhancing Australia's defence capabilities. (Photo by Kerrie Dougherty)

However, unlike many other nations, but in line with its approach to civil space activities, Australia has not developed its own national military satellite systems, but has relied instead on access to these systems being provided by its major alliance partners, the United Kingdom and the United States. This space access, and the expertise to exploit it for the country's security and defence, is considered an integral component of the Australian Government's critical infrastructure, and has represented the nation's most significant investment in space activities since the end of the Joint Project.

To provide maximum national value from the space-based assets to which Australia has access, the Defence Science and Technology Organisation – which in 2015 was re-named Defence Science and Technology Group (DST Group) – has, since its formation, undertaken a wide range of applied research and technology innovation in the operation of systems to access and exploit military and civil satellite communications, remote sensing and position, navigation and timing products utilising the GPS system.

In particular, DSTO gained significant expertise in developing and operating military satellite communications systems. At the beginning of the Twenty First Century, this expertise was applied to the development of the Defence communication payload that was carried aboard the Optus C1 communications satellite, launched in 2003. Australia's first dual use communications satellite, Optus C1 was partly funded by the ADF and was, at the time of launch, the world's largest hybrid commercial and military satellite.

Australia's ongoing overseas military engagements in the wake of the 11 September 2001 terrorist attacks in the United States have highlighted the fact that reliance on allied space systems means that ADF requirements for space access may, at times, have a lower priority than the requirements of the host nation. Consequently, with its ever-increasing use of space systems to underpin its modern digital network-based warfighting capabilities, in 2007 Australia agreed to fund the sixth satellite in the United States' Wideband Global SATCOM network, to ensure its access to global military communications services.

The WGS 6 satellite was launched from Cape Canaveral in August 2013, becoming Australia's first nationally-owned defence satellite.

CUBESATS FOR DEFENCE RESEARCH

One of the breakthrough technologies that have enabled the NewSpace, or Space 2.0, movement (see Chapter 8), tiny cubesats, which are cheap to build and launch compared to traditional satellites, are ideal testbeds for the development and space qualification of new space instruments and technological innovations. Advances in small low-cost space platforms like cubesats provide opportunities to enhance Australian Defence Force capabilities and also revitalise Australian space research. In 2017, two cubesats carrying defence science experimental payloads in which the Defence Science and Technology Group has participated, are being launched.

Biarri

The Biarri project is being jointly undertaken by the 'Five Eyes' partners (see below), Australia, New Zealand, Canada, the United Kingdom and the United States to explore cubesat formation flying and better understand the drag and lift forces experienced by cubesats. The Australian contribution, led by BAe Systems, in collaboration with DST Group, the Australian Centre for Space Engineering Research (ACSER) and Electro Optic Systems (EOS), will also verify the performance of Australian-developed Namuru V32R3A GPS receivers and EOS techniques for space situational awareness.

The Biarri program consists of two stages. The first, Biarri Point, employs a single cubesat, flown as a risk mitigation flight, before the Biarri Squad constellation of three satellites, which will carry out the main experimental program. The Biarri Point satellite, developed by the US, uses a cubesat bus (body) provided for military technology demonstration missions by the US National Reconnaissance Office. Its payload includes the Namuru GPS receiver developed by ACSER, University of New South Wales (Sydney), in partnership with Defence Science and Technology Group. It is also fitted with corner reflectors, which allow the satellite to be located using satellite laser ranging: it will be tracked by EOS' laser tracking facility on Mount Stromlo, Canberra.

Along with three other Australian cubesats (further discussed in Chapter 8), Biarri Point was launched to the International Space Station on 19 April. It was successfully released into orbit from the space station a few weeks later.

Buccaneer

Buccaneer is a joint project of the University of New South Wales (Canberra) and the Defence Science and Technology Group, designed to use cubesats to calibrate the Jindalee Operational Radar Network (JORN) that monitors Australia's air and maritime regions to the north and west of the country. The Buccaneer cubesats will receive signals from JORN, using a high frequency receiver, that will then be processed and analysed to improve the calibration of the JORN system.

While the spacecraft bus has been purchased from a commercial cubesat manufacturer in the US, DST Group is providing the digital High Frequency (HF) receiver payload, while UNSW Canberra is contributing a payload camera that will be used to confirm correct deployment of the HF antenna. UNSW Canberra will also perform photometry experiments, telescopically observing Buccaneer in order to better understand how measurements of the light signal reflected from space objects can be used to infer information about how they are tumbling in space.

Like Biarri, the Buccaneer project consists of two phases – a risk mitigation satellite, followed by an operational satellite. The risk mitigation satellite, which will test the deployment of the HF antenna system, is expected to be launched, courtesy of the United States, in July/August 2017. If successful, the operational mission is expected to follow in 2018. The program will be supported by ground stations at DST Group's South Australian facility and at UNSW Canberra.

SPIES IN THE SKY

During the Cold War, the majority of satellites launched each year were for military and national security purposes. Dedicated military satellite networks were developed to support warfighting and defence activities and obtain intelligence considered vital to safeguarding national security. While improvements in satellite technologies and systems have gradually increased satellites' capabilities and lifespans in orbit, reducing the need for large numbers of military satellite launches, such launches still make up a significant proportion of global space activity each year.

Since the beginning of the Space Age, surveillance or 'spy' satellites have been employed by many nations to gather intelligence about their opponents. These satellites include both photographic reconnaissance satellites and radar/infra-red satellites, which can detect such things as concealed facilities and submarines and provide early warning of missile launches. Throughout the Cold War, the resolving power of photoreconnaissance satellites was much greater than that of civilian remote-sensing satellites, although the rapid development of digital imaging technologies in the past two decades has now placed equivalent capabilities in the hands of

Above Australia's first defence research cubesat, Biarri Point, designed to test Australian-developed GPS technology and satellite laser tracking. (Photo courtesy of Defence Science and Technology Group)

ABOVE *The Buccaneer defence research cubesat being tested in a laboratory at DST Group, Edinburgh in South Australia. (Photo courtesy of Defence Science and Technology Group)*

Space Situational Awareness

As the number of objects in Earth orbit grows, ranging from large operational satellites and spent rocket stages, to tiny flecks of paint and other spacecraft debris, space situational awareness has become a matter of particular concern in both the defence and civil space sectors, due to the ever-increasing military and commercial reliance on space-based assets. The term Space Situational Awareness (SSA) refers to the ability to monitor and predict the physical location of natural and manmade objects in Earth orbit, so as to avoid collisions that may damage, disable or destroy spacecraft, or create further space debris.

Australian company Electro Optic Systems has become a global leader in SSA through its expertise in satellite laser tracking systems, which bounce lasers from dedicated reflectors attached to a satellite back to Earth. This technique enables a precise location for the satellite to be determined. EOS has also been at the forefront of efforts to mitigate space debris hazards by experimenting with the use of lasers to abrade the surface of debris items, increasing their friction and thus causing them to eventually slow down enough so that they will re-enter and burn up in the atmosphere.

commercial companies, and even casual browsers of Google Earth. Signals intelligence (Sigint) or Electronic intelligence satellites (Elints or Ferrets) listen in to radio, telephone, data and telemetry channels to discover political and military information and also to determine the capabilities of other nations' weapons systems.

Just like civilian satellite networks, military space systems require ground stations to carry out satellite command and control operations and receive data from their intelligence gathering activities. Just as Australia's geographical position made it an ideal location for NASA tracking stations, so too has it been suitable for military and intelligence satellite ground stations.

Australia's political alliance with the United States meant that it was approached in the early 1960s to host tracking stations in connection with the Transit satellite navigation and positioning system, which was used not only to improve navigational accuracy for US naval ships and submarines, but also to produce better guidance and targeting data for its ICBMs. The first Transit tracking station in Australia was established at Smithfield, a suburb of Sydney, in 1961. It remained in operation there for ongoing satellite geodesy programs until 1994, when it was moved to DSTO in Salisbury.

Bases for Debate

Three major ground stations for the US intelligence-gathering network were established in Australia between 1967 and 1970. Shrouded in secrecy, and agreed to by Government without public discussion or acknowledgement, these three facilities concealed and obscured their functions behind euphemistic designations. They became the subject of national debate and protest throughout the Cold War, on the basis of public belief that they operated without Australian oversight or control and that their presence in Australia invited the possibility of nuclear attack, in the event that the Cold War turned hot.

Free Space

When Sputnik 1 was launched in 1957, there were no claims by other nations that its overflight was a breach of territory. This eventually led to the acceptance of the 'non-appropriation' principle of international law, which was enshrined in the Outer Space Treaty of 1967. This means that satellites may travel over whatever countries their orbits carry them without those countries claiming that their 'territorial airspace' has been infringed. Because of the international acceptance of this principle, intelligence gathering satellites are tolerated and not removed by politically opposing countries (at least in peacetime).

Checking up on political opponents from above, surveillance satellites, euphemistically referred to as 'national technical means of verification', have played an important role in verifying adherence to the arms control treaties, which have made disarmament is possible. Consequently, during the Cold War, the intelligence provided by surveillance satellites made an important contribution to the maintenance of the balance of power between the Superpowers, contributing significantly to the prevention of nuclear warfare since 1945.

Harold E Holt Naval Communications Station, North West Cape

The first of the major US space installations to be established in Australia was the North West Cape Naval Communications Station, near Exmouth, Western Australia. Opened in 1967 to provide VLF radio communication with US nuclear ballistic missile submarines, the base was renamed the Harold E Holt Naval Communications Station in 1968, after the Australian Prime Minister who disappeared in a swimming accident that year. By the early 1970s, a Defence Satellite Communications System (DSCS) ground station had been established for communication with the US surface fleet and attack submarines – a war-fighting role rather than a deterrent one, making the base a potentially high priority Soviet nuclear target.

Initially, the Australian Government had no control over or access to the contents of the communications passing through North West Cape and it was not until the 1980s that 'joint' operation of the station really commenced. By 1992, the US presence was diminishing and Australia took over responsibility for the facility in 1997, although US involvement and funding continues: in 2008, Australia and the US signed an agreement on future joint use of the facility for another 25 years. In 2013, it was announced that the Space Surveillance Telescope (SST), a component of the US Space Surveillance Network, would be transferred to the station and become operational in 2020. A C-Band Space Surveillance Radar is also being installed, to be operated remotely by the RAAF. It will provide a Space Situational Awareness capability, allowing the tracking of space assets and debris.

Joint Defence Facility Nurrungar

The Joint Defence Space Communications Station at Nurrungar, later renamed the Joint Defence Facility Nurrungar, was located about 10km from the Woomera Village and commenced operation in 1970. During the Cold War, Nurrungar was of crucial importance to the United States, since it provided surveillance of USSR ICBM launches: it was, therefore, considered a prime Soviet target. Under the control of the US Air Force and Australian Department of Defence, Nurrungar monitored US Defence Support Program early warning satellites, which used infra-red sensors to detect the launch of ICBMs and nuclear weapons tests. During the 1990-91 Gulf War, it also provided monitoring of Scud missile launches from Iraq. After the end of the Joint Project, staff from Nurrungar made up a sizeable proportion of the Woomera Village's inhabitants, until it, too, was closed in 1999, when its operations were moved to the Joint Defence Facility at Pine Gap.

The Australian Government and Defence Forces had full access to Nurrungar's data, but the facility was nevertheless particularly controversial. Although its early warning and nuclear detonation detection roles were important safeguards against surprise attack or accidental nuclear conflict during the Cold War, its capabilities could also make a critical contribution to US nuclear war-fighting capabilities.

Joint Defence Facility, Pine Gap

Although North West Cape and Nurrungar could arguably have been sited outside Australia, the United States' most important intelligence gathering ground station in the country needed to be located in Australia because of the particular orbits of the satellites it controlled. These strategically significant space assets provide coverage of about one-third of the world, including parts of the Soviet Union/Russia, China and the Middle East. Pine Gap's remote location, about 18km from Alice Springs, was selected to protect its own communications back to the United States from interception by Soviet surveillance ships off the Australian coast.

Established as the Joint Defence Space Research Facility, Pine Gap, in 1970, the facility's name was changed to the Joint Defence Facility, Pine Gap, in 1988. One of the largest satellite ground stations in the world, Pine Gap is jointly managed by the Central Intelligence Agency (CIA), the US National Security Agency (NSA) and the US National Reconnaissance Office (NRO). During the Cold War, the Signals Intelligence captured by the station played an essential role in the verification of some major arms control agreements. While Australia did not initially have access to some of the data passing through the station, since 1980 the operation could be more truly described as 'joint'.

In the post-Cold War era, Pine Gap is a key component of the ECHELON global surveillance network, a Sigint surveillance program operated by the 'Five Eyes'. Five Eyes is an intelligence alliance comprising the United States, Australia, Canada, New Zealand and the United Kingdom: these countries are bound by the UKUSA Agreement, a treaty for joint co-operation in signals intelligence. Created in the late 1960s to monitor the military and diplomatic communications of the United States' Cold War adversaries, it is controversially claimed that ECHELON has now evolved beyond legitimate strategic intelligence gathering to also become a global system for the interception of private and commercial communications across the digital domain.

In addition to its Sigint function, Pine Gap also carries out early warning surveillance (taken over from Nurrungar). Another important and controversial function is its role in locating radio signals, identifying targets for the US drone program, which supports military operations in Afghanistan, Iraq and Syria, in which both the United States and Australia are involved.

AUSTRALIAN SIGINT SPACE FACILITIES

In addition to the US bases it accommodates, Australia operates its own regionally important Sigint program, with satellite interception facilities under the control of the Australian Signals Directorate (ASD) (known as the Defence Signals Directorate from 1977-2013). The ASD provides foreign signals intelligence to the Australian Defence Force and Australian Government to support military and strategic decision-making. It is primarily concerned with obtaining intelligence relevant to Australia's regional security interests in South-East Asia and China.

RIGHT A Defence Support Program satellite being launched from the Space Shuttle cargo bay during the STS 44 mission in November 1991. DSP satellites were monitored via the Nurrungar station for early warning of missile launches. (Photo courtesy of NASA)

The ASD operates three ground stations: the Shoal Bay Receiving Station, located at Shoal Bay, near Darwin; a small station on the Cocos (Keeling) Islands; and its major facility at Kojarena near Geraldton, Western Australia. It also has personnel stationed at Pine Gap. Shoal Bay and Kojarena have been identified as stations within the ECHELON network. The Australian Defence Satellite Communication Station (ADSCS) at Kojarena has four satellite tracking antennae that intercept communications from regional satellites of interest, as well as from international communications satellites. Kojarena provides Australian access to the WGS system, including control of Australia's WGS satellite, and hosts the five antennae that form the joint US-Australian ground station for the US Mobile User Objective System (MUOS), a narrow-band networked satellite constellation that enables secure 3G mobile telecommunications. The Kojarena facility is one of four MUOS ground stations.

Australia's utilisation of space for defence and national security purposes has been its longest and most continuous space activity. While Australia may be said to have gained no benefit from the early long-range missile research conducted at Woomera, the presence of US intelligence gathering satellite ground stations in the country contributed to the global strategic balance and arms control during the Cold War, though arguably leaving Australia open to nuclear attack. In the post-Cold War strategic environment, Australian Governments have given bi-partisan support to defence space research contributing to the Australian Defence Force's increasing reliance on space-based services, while also embracing the use of the high ground of space for national security through participation in US-led global surveillance programs. It is perhaps only to be lamented that these space activities have been the products of defence and foreign affairs policies, rather than a component of a national space policy.

Geospatial Intelligence

As demand for satellite-provided remote sensing and environmental data increased across the government sector in the latter half of the 1990s, the Defence Imagery and Geospatial Organisation (DIGO) was created in 2000. Its role was to provide geospatial intelligence derived from satellite imagery and other sources in support of Australia's defence and national interests.

Renamed the Australian Geospatial-Intelligence Organisation (AGO) in 2013, this agency obtains geospatial and imagery intelligence to: support ADF operational, targeting, training and exercise requirements; support Commonwealth and State authorities in carrying out national security functions; and provide imagery, unclassified geospatial products, technical assistance and support to emergency service agencies for carrying out their emergency response functions. AGO provides assistance to the ADF in support of military operations and cooperates with the Defence Forces on intelligence matters. It operates from two sites: one in Canberra, and the other in Bendigo, Victoria.

BELOW The Joint Defence Facility, Pine Gap, is the largest and most important US intelligence gathering facility in Australia. (© Commonwealth of Australia)

Defence Hypersonic Research in the Twenty-First Century

Where the early hypersonic research at Woomera was aimed at developing missile re-entry vehicles and advanced military aircraft, hypersonic research at Woomera in the Twenty-First Century has been focused on the development of SCRAMJET technology, which offers the possibility of hypersonic sub-orbital intercontinental transport and cheap access to space – both desirable for defence and commercial space activities.

The SCRAMJET, or supersonic combustion ramjet, is a form of jet engine that eliminates the need for heavy turbines by travelling at speeds that force air through the engine at supersonic velocities, to combine with fuel and produce thrust. The nature of their design means that SCRAMJET engines operate at speeds Mach 5 and Mach 20, making them well-suited to propel sub-orbital aerospace planes or form the upper stages of launch vehicles. Australia has been a world leader in SCRAMJET research since the 1980s, as outlined in Chapters 7 and 8.

HyShot

Building on its experience in hypersonic SCRAMJET research, the University of Queensland's Centre for Hypersonics initiated the HyShot Program to provide experimental correlation between pressure measurements made of supersonic combustion in its hypersonic shock tunnel and those actually observed in the flight of a prototype SCRAMJET engine.

Because of the dual defence and civilian applications of SCRAMJET technology, HyShot drew supporting sponsorship and technical collaboration from a wide range of international agencies and industrial partners. These included: the University of Queensland, Astrotech Space Operations, Defence Evaluation and Research Agency (DERA, UK, which was privatised as QinetiQ in 2002), NASA, DSTO, the Australian Department of Defence, the then-Department of Industry Science and Resources, DLR (the German Aerospace Centre), the Seoul National University (Korea), The Australian Research Council, ASRI, Alesi Technologies (Australia), the National Aerospace Laboratories (NAL, Japan), NQEA, ARDU, the US Air Force Office of Scientific Research (AFOSR) and Luxfer, Australia.

Between 2001 and 2007, five test flights were carried out at Woomera, during which SCRAMJET prototypes provided by both the Centre for Hypersonics and the British defence company QinetiQ were flown. The first four flights launched their SCRAMJET payload using a two-stage Terrier-Orion sounding rocket to accelerate the engine to the hypersonic velocity required for supersonic combustion to take place. Following a ballistic trajectory similar to that used for the Black Knight and SPARTA hypersonic research, the launch vehicle lifted the test engine to an altitude of around 330km. As the spent second stage and its attached SCRAMJET payload fell back toward the ground, it reached Mach 7.6 between 35km and 23km. At this point, the payload was ignited for a period of 6–10 seconds, during which the measurements of the combustion pressures were taken, although the system was not designed to produce actual thrust.

The first HyShot experiment, launched on 30 October 2001, was unsuccessful, due to a launch vehicle failure, but HyShot II, on 30 July 2002, was carried out flawlessly and is generally considered to be the world's first successful flight of a SCRAMJET engine, demonstrating that the technology was viable.

Following this successful flight, the South Australian and Queensland Governments, the Australian National University, the University of New South

Defence Hypersonic Research in the Twenty-First Century

Above A sounding rocket streaks skyward carrying a University of Queensland experimental SCRAMJET during the HIFiRE 5b test. (Photo courtesy of Defence Science and Technology Group)

Wales' Australian Defence Force Academy campus in Canberra and the University of Queensland, combined to establish the Australian Hypersonics Initiative, intended to co-ordinate and exploit research in hypersonics and SCRAMJET technology and boost Australia's potential to become a leading participant in the field of air-breathing launchers and high-speed missiles.

HyShot III, which took place on 25 March 2006, repeated the HyShot II flight profile using a prototype SCRAMJET provided by QinetiQ, which was of a different design to the Queensland engine. HyShot IV followed only days later, on 30 March, with another Queensland engine. However, despite a successful launch, no data was received from the engine itself.

The final launch in the HyShot program was a follow-on project dubbed HyCAUSE (Hypersonic Collaborative Australian/United States Experiment), a collaboration between the US Defence Advanced Research Projects Agency (the successor to ARPA which had participated in the early hypersonic re-entry programs at Woomera) and DSTO, representing the research collaborators in the Australian Hypersonics Initiative. Using a Talos-Castor two-stage vehicle, HyCAUSE was launched on 15 June 2007, achieving a velocity of Mach 10 during the engine test phase.

HIFiRE

In 2006, DSTO and US Air Force Research Laboratory (AFRL) entered an agreement to advance research in hypersonic flight, building on the success of the HyShot program. This resulted in the HyCAUSE launch of 2007 and the establishment of a further program, the Hypersonic International Flight Research Experimentation (HIFiRE). One of the largest collaborative ventures between DSTO and the United States, HIFiRE partners also include the University of Queensland and University of New South Wales at the Australian Defence Force Academy, as well as the Boeing company.

The objective of the program is to investigate the fundamental science of hypersonics technology and its potential for next generation aerospace systems, such as hypersonic 'quick strike' weapons and 'quick reaction' space launchers. Data from the first HIFiRE tests was incorporated into the development of Boeing's X-51 Waverider SCRAMJET demonstrator, which was successfully test flown in May 2013.

The first hypersonic test under the HIFiRE program (HiFIRE 0) took place at Woomera in May 2009, with the launcher reaching an altitude of 200km before releasing its test vehicle to dive back into the atmosphere. Nitrogen thrusters manoeuvred the test vehicle in space, positioning it correctly for re-entry. Even before this first launch, the HIFiRE collaboration had already achieved some significant milestones in the design, assembly and pre-flight testing of hypersonic vehicles and the design of complex avionics and flight systems.

The second HIFiRE flight (HIFiRE 1) took place at Woomera in March 2010, with the goal of measuring hypersonic boundary-layer transition and shock boundary layer interactions in flight. Subsequent HiFIRE tests took place outside Australia and in 2012 the HIFiRE program was awarded the prestigious von Karman Award for International Cooperation in Aeronautics by the International Council of the Aeronautical Sciences.

In 2016, the HIFiRE program returned to Woomera, with a University of Queensland SCRAMJET flight successfully carried out on 18 May, as HIFiRE 5b. Further HIFiRE trials are planned for Woomera in coming years and will include tests of a SCRAMJET-powered waveriding vehicle, designed by the Centre for Hypersonics.

The Defence potential of SCRAMJET propulsion was particularly acknowledged by its inclusion as a research area requiring government support in the Defence White Paper of 2016.

BELOW *HyShot IV SCRAMJET test ready for launch in March 2006. The heavy-duty launcher was originally installed by Japan for its NEXST supersonic flight model tests. (© Commonwealth of Australia, Department of Defence 2015)*

ABOVE The first complete image of Australia seen from orbit. It was captured by NASA's second applications technology demonstrator satellite ATS 2 and received at Cooby Creek tracking station in April 1967. (Photo by Neil Sandford. Scan by Colin Mackellar. Courtesy of www.honeysucklecreek.net)

LEFT The first three generations of INTELSAT satellites, with INTELSAT 1 on the right. Australia was first connected to the world via INTELSAT II and III. (Photo courtesy of NASA)

CHAPTER 6

An Infrastructure in Orbit: Satellites at Your Service

When flicking a light switch or turning a tap, rarely do we consider the infrastructure that provides power and water, so deeply does it underlie daily life. Space-based services provided via satellite have equally become so seamlessly integrated into the modern world that few people are aware of the powerful global infrastructure in orbit that helps to support it. In almost every aspect of everyday life, space technology plays a role: facilitating online shopping, providing directions to a new location, putting food on the table, watching the news or being entertained, deciding what to wear for the day — somewhere behind these and many other aspects of daily life, there is a satellite at your service.

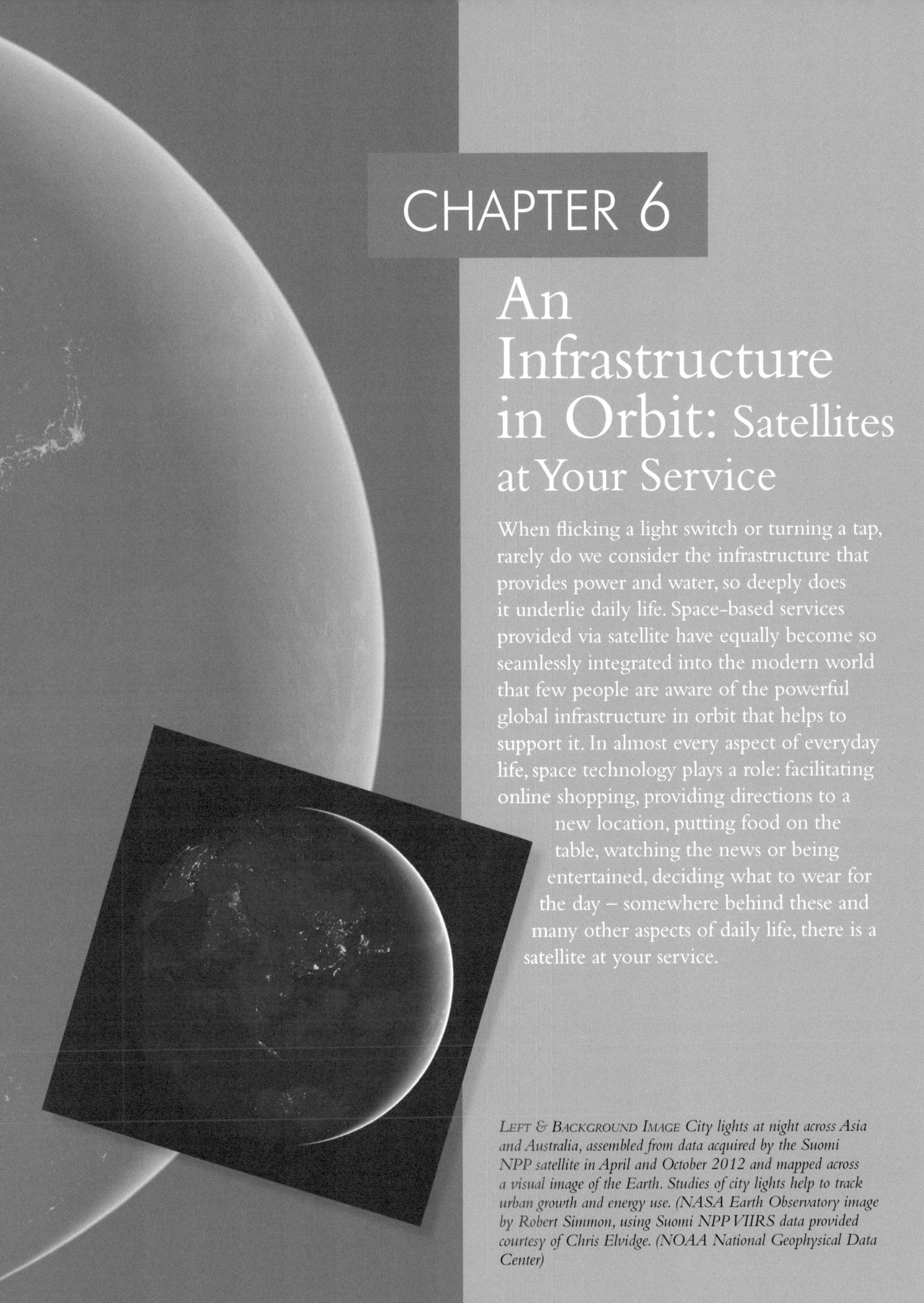

LEFT & BACKGROUND IMAGE City lights at night across Asia and Australia, assembled from data acquired by the Suomi NPP satellite in April and October 2012 and mapped across a visual image of the Earth. Studies of city lights help to track urban growth and energy use. (NASA Earth Observatory image by Robert Simmon, using Suomi NPP VIIRS data provided courtesy of Chris Elvidge. (NOAA National Geophysical Data Center)

Satellite Orbits

Satellites can travel in many different orbits around the Earth, but most applications satellites can be found in three types of orbits:

Geostationary orbit (GEO), approximately 36000 kilometres above the Earth's equator, is where satellites travel at a speed that is equal to that of the planet's rotation. Consequently, each satellite remains at a fixed point above the equator. A satellite in geostationary orbit can observe about one-third of the Earth's surface, so only three satellites are needed, in any network, to provide world-wide coverage. The majority of the world's communication satellites are in geostationary orbit, which is also used by meteorological and environmental satellites.

Polar orbit, as its name indicates, is an orbit that travels over the Earth's poles. As the Earth rotates below it, a satellite in polar orbit will pass over the same place on the Earth again after a certain number of days. The return visit time depends upon the height at which the satellite orbits. Many remote sensing satellites travel in polar orbits.

Low Earth orbit (LEO), relatively close to the Earth at only a few hundred kilometres, was the first orbit used by satellites. It is still the region where human space activities occur, below the dangerous van Allen radiation belts. The International Space Station and the Hubble Space Telescope are in LEO, as are many scientific satellites.

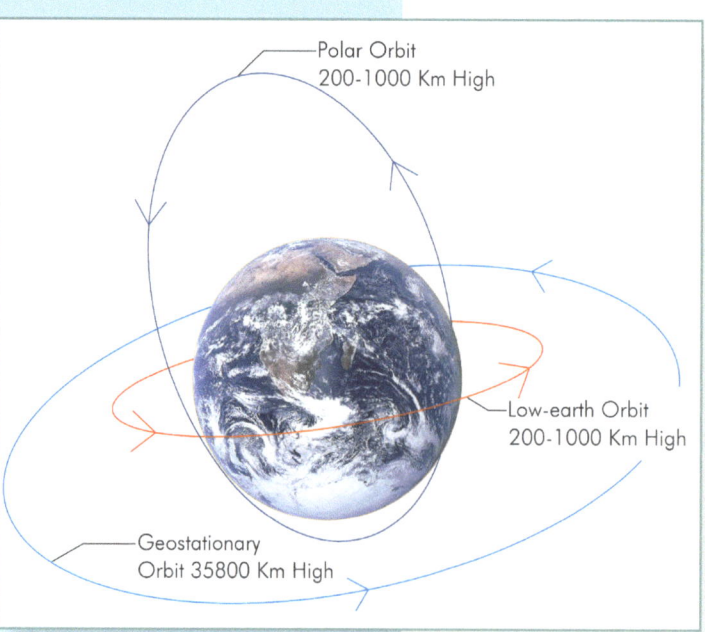

Australia was an early adopter of space-based satellite applications, making use of weather and communications satellites as soon as they were available to the country in the 1960s. Today, the nation is one of the heaviest users of satellite services, to assist in overcoming the 'tyranny of distance' and managing the resources and environment of a vast continent with a small population.

Linked by invisible threads – Communications Satellites

'Live via satellite': once every television news program made this statement proudly when it carried a direct report from overseas. Today, when multiple platforms allow instant 24-hour access to news, information and entertainment from around the world, and we can effectively call anyone, anywhere, any time, the role that satellites play in delivering content and communication is taken for granted. Few people realise that international telecommunications via satellite was the first space application to have an impact on everyday life – and the first to be profitably commercialised.

The first commercial communications satellite, Telstar, entered service in 1962, barely five years after the launch of Sputnik 1, linking North America and Europe. Although Telstar could only handle one television channel or a handful of telephone calls at a time, the satellite represented the beginning of a revolution in how we communicate around the world – a revolution that still continues today in our ever-more digitally connected lives.

Australia became connected to the world via satellite in 1966, with the development of INTELSAT, the world's first global satellite network. Just as the earlier communication technologies of telegraph, telephone and radio had gradually reduced

Australians' isolation from each other, scattered across the continent, and from the rest of the world, access to international telecommunication via satellite allowed governments, businesses and everyday people to exchange information, connect with colleagues, improve the speed of business transactions or keep in touch with loved ones overseas with more immediacy than ever before.

INTELSAT AND INMARSAT: CONNECTING AUSTRALIA TO THE WORLD VIA SATELLITE

Once the first communications satellites proved the financial viability of orbital-based services, moves were quickly made to establish a global satellite telecommunications network. The International Telecommunications Satellite Organisation, known as INTELSAT, was established in 1964 to create the world's first geostationary satellite network. INTELSAT was an international, intergovernmental consortium, made up of member nations that were represented by their national postal and telecommunications authority. Member nations contributed to the cost of establishing, operating and maintaining the satellite system, but received a return for that investment through the revenue that was generated from satellite usage fees. INTELSAT's services were open to any nation to use, and all users paid the same rates, a policy that helped to bring satellite communications within economic reach of developing nations.

Recognising the value of satellite communications to the nation, Australia was one of the eleven founding members of the INTELSAT consortium and would become its sixth largest shareholder. Australia's interests in INTELSAT were represented by the Overseas Telecommunications Commission (OTC), renamed the Overseas Telecommunications Corporation in 1989. OTC was created by the Australian Government in 1946 as the national body responsible for the management and development of Australia's public telecommunication system – which at that time meant radio, telegraph and cable. As it developed growing expertise in satellite operations during the 1960s and 70s, OTC would come to play a significant role in INTELSAT's advisory committees for satellite procurement and ground station development.

Although INTELSAT launched its first satellite, INTELSAT 1 (nicknamed Early Bird), in April 1965, Australia did not have access to the system until the Intelsat II network was established in late 1966-67 with satellites over the Indian and Pacific Oceans. The country's first satellite ground station, or Satellite Earth Station, was built at Carnarvon in Western Australia. This location was selected at the request of NASA, in order to provide more reliable and immediate communications between its Carnarvon tracking station and the United States during space missions: a lightning strike had destroyed the telephone lines south from Carnarvon during the Gemini 1 unmanned test flight, requiring complicated work around using old telegraph lines in order to get information to and from the tracking station. Through the Carnarvon Earth Station, signals from the tracking station could be sent directly via satellite to NASA facilities in the US.

BELOW Modern aerial view of the OTC Carnarvon Satellite Earth Station, which is now a museum. The original antenna, used for the first satellite broadcast, NASA communications and later for TTC&M work is on the upper right; the 30m dish antenna was constructed in 1969. (Author's collection)

Above The unusual shape of the original cassegrain feed-horn antenna earned it the nickname 'the sugar scoop'. (Photo courtesy of Guntis Berzins and www.honeysucklecreek.net)

Australia's first satellite broadcast occurred on 25 November 1966, when 'Down Under Comes up Live' linked Carnarvon with the UK (see below for more on this unique story). At the time, it was the world's longest distance satellite broadcast. Regular television broadcasts, however, did not commence until OTC established a new Earth Station at Moree, NSW, in 1968, linking to the Intelsat II Pacific Ocean satellite. This connection to the world via satellite was considered so important that it was represented on an Australian stamp in 1968. Another station in Ceduna, South Australia, opened in 1969, linking to the new Intelsat III Indian Ocean satellite. At this early period, the technology required the use of large antennae, which had to be located in regions away from the electromagnetic interference of cities, in order to successfully receive the relatively low-powered signals from the satellites.

In late 1968, OTC was awarded an INTELSAT Tracking, Telemetry, Command and Monitoring (TTC&M) contract and the original Carnarvon antenna was upgraded for this role. TTC&M stations do not simply transmit and receive signals from communications satellites, rather they provide the important 'housekeeping' functions necessary to maintain a satellite network: tracking satellites, keeping check upon their operational health and commanding and controlling their operations in orbit. OTC would continue to receive TTC&M contracts from INTELSAT over the following decades, allowing it to develop considerable experience and expertise in this area of satellite operations.

When improvements in satellite technology enabled ground stations to be located closer to cities, OTC closed its original stations and established new facilities: at Healesville, near Melbourne, in 1984 and, in 1987, at Gnangara, near Perth, Western Australia and Oxford Falls, near Sydney. The Ceduna station was donated to the University of Tasmania for use as a radio astronomy observatory. In the late 1980s, OTC also installed ground stations at Australia's Antarctic bases, allowing them to maintain direct contact with Australia and the rest of the world. By the 1990s, INTELSAT satellites were providing approximately 70 per cent of Australia's international telephone and television communications.

In 1979, a new international satellite consortium, the International Maritime Satellite Organisation (INMARSAT) was created to establish a satellite communications network for maritime service. Structured and operated as an intergovernmental body similar to INTELSAT, INMARSAT's role quickly expanded to include aeronautical communications. By the 1990s, the organisation was moving into mobile services for vehicles and a search and rescue service. With Australia's reliance on maritime trade and aviation for international travel and commerce, it became a founding member of the new organisation, with OTC again as the national representative. When OTC's Perth International Telecommunications Centre at Gnagara opened in 1986, it included the first Pacific and Indian Ocean region INMARSAT ground station in the southern hemisphere.

Until 1992, OTC was responsible for all telecommunications carried by INTELSAT and INMARSAT satellites into and out of Australia. Following the reorganisation of Australian telecommunications that occurred in 1992, access to these satellites became available to both OTC and the new telecommunications carrier Optus, created by the privatisation of the Aussat domestic satellite system.

Aussat: Australia's National Satellite System

In the 1970s, many nations began to develop domestic communications satellites to serve their intra-national requirements: Canada established the world's first domestic satellite system with the launch of its Anik A1 satellite in 1972, while Indonesia became the first developing nation to establish a domestic system, when its first Palapa satellite was placed in orbit in 1976.

As early as 1969, the Postmaster General's Department (PMG) had carried out experiments using NASA's ATS 1 satellite, via the Cooby Creek tracking station, to investigate the use of satellites for telephone services to remote outstations across Australia. However, the first suggestion for an Australian domestic satellite system came in a 1977 report to the Fraser Government commissioned by Australian media mogul Kerry Packer. Despite advice from Telecom Australia (the national phone service provider and successor to the PMG) that there was not an economic case to justify a national satellite system, the Government established a task force to inquire into the concept: its report to Parliament in 1978 recommended that planning for a government-owned Australian satellite system should commence immediately. In 1979, a Canadian communications satellite took part in experimental transmissions to remote areas in New South Wales and Queensland to investigate the technological means of introducing satellite broadcasting direct to homes in the outback.

The decision to establish a domestic satellite system was not without its dissenters: it was opposed by some sections of the government bureaucracy on financial grounds, while Telecom Australia, with its own interests in mind, suggested that expansion of its own terrestrial communications system was a viable alternative. However, the project which came to be called Aussat was conceived in an era when national satellite communications systems were becoming an international status symbol. The support of powerful media interests, that saw a domestic satellite system as a way to expand their broadcasting networks across the country, also undoubtedly played a role in swaying the Government to proceed with the project. Even though the system was financially marginal at best, and would eventually require direct support grants, its proponents would later insist that telecommunications policy actually prevented the system from operating as it was designed to do – in competition with Telecom Australia's domestic services.

Established under the 1984 Satellite Communications Act, Aussat Pty Ltd was initially 75 per cent owned by the Commonwealth of Australia, with the remainder owned by Telecom Australia. Aussat's establishment cost of $300 million was largely consumed by the cost of purchasing three satellites from the US, two satellite launches using the Space Shuttle in 1985 and a third launch on an Ariane rocket in 1987, and the construction of ground stations. With a staff of about 300, many drawn from the ranks of experienced satellite operations personnel at OTC, Aussat had the highest concentration of satellite communications and space engineering experience in Australia at the time.

Above The first Aussat communications satellite, launched from the Space Shuttle Discovery during the STS 51I mission in August 1985. (Photo courtesy of NASA)

Aussat constructed eight major city ground stations providing for single point or multipoint contact across Australia, New Zealand and Papua New Guinea, which were also covered by the planned satellites' broadcasting footprint. The main control station was built at Belrose in Sydney, with a back-up at Perth, both providing TTC&M services. The other city stations were automated and not normally staffed. Major system uses included television and radio program interchange and distribution by major media networks, separate corporate and government communications networks, aeronautical services and Australian Defence Force communications: by the early 1990s there were more than 5000 ground terminals connected to Aussat's services.

To provide the public good benefits, which had been one of the rationales for its establishment, Aussat developed a range of small, low-capacity, fixed and mobile ground stations for use by people living in remote areas. These allowed access to the Homestead and Community Broadcast Satellite Service (HACBSS), which was designed to improve the delivery of distance education services in rural and remote areas and Aboriginal communities: it also provided television services to many parts of the outback for the first time. The analogue HACBSS was replaced in 1998 by Optus Aurora, a free-to-view service that provided television and radio services to remote areas (and reception 'black spots' in other parts of the country) via Optus satellites. The Optus Aurora service was itself replaced by the Viewer Access Satellite Television (VAST) between 2010 and 2013, which provides access to a full range of digital channels.

Changes in the Air: Privatisation in Australia and Overseas

The satellite telecommunications landscape in Australia changed markedly in the 1990s, with policies of 'economic rationalism' encouraging the privatisation of many government-owned entities. In 1992, Telecom Australia and OTC were merged into the Australian and Overseas Telecommunications Corporation (AOTC), which adopted the name Telstra for the corporation in 1995. Telstra was gradually partially privatised beginning in 1997. Aussat was fully privatised in 1992 and sold to Optus Communications (the name means 'choice' in Latin) to become a competitor to Telstra. Originally named Optus Communications Pty Ltd, the company was initially 51 per cent owned by an Australian consortium of business and insurance companies, with the other 49 percent owned by two foreign telecommunications companies, Bell South from the US and Cable and Wireless from the UK. Behind the trading name Optus, the company has changed its name and ownership several times, due to business buyouts and takeovers over the past 25 years.

Imparja Television: Indigenous Broadcasting

The remote area television broadcasting coverage provided by the Aussat system enabled the creation of Australia's first Aboriginal-owned television network, Imparja Television. Based in Alice Springs, Imparja (an Arrernte word meaning 'footprints') commenced operation in 1988, after the Central Australian Aboriginal Media Association, already operating an Indigenous radio station, was awarded a commercial television licence to broadcast across Central Australia. The company's aim was to serve the Arrernte people wherever they lived across the central regions of Australia. At the beginning of her tenure, Imparja Television's first Chair, Freda Glynn, was the only female Chair of any television network in the world.

In addition to carrying commercial content from the major national networks, the ABC and SBS, Imparja began producing its own material in 1994, with the introduction of the children's program Yamba's Playtime, the first Indigenous-themed preschool program: it featured the network's official mascot, 'Yamba the Yerrampe' (honey ant). In the early 2000s, the 'Imparja Info Channel' was launched, providing programming, news, and community information to remote Aboriginal communities. In 2007, this channel was replaced by National Indigenous Television (NITV). Through its services and productions, Imparja has played an important role in bringing about an increased visibility of Aboriginal identity in the Australian media.

Optus began construction of new ground stations in Sydney and Perth. The Aussat satellites, now renamed 'Optus', became a back-up for the national optic-fibre based system that the company began to roll out, while continuing to provide broadcasting and remote area services, together with new satellite services to mobile vehicles, to complement existing terrestrial cellular mobile telephone systems.

Optus also acquired the second generation satellites that Aussat had ordered before its privatisation, the first of which, Optus B1, was launched in March 1992. Optus B2, however, was lost in a launch accident later that year. Optus had become one of the first Western companies to launch on the Chinese Long March 2e vehicle, which was briefly offered as a commercial launcher in the early 1990s, before US restrictions, prohibiting the launch of any satellite incorporating US-made components on Chinese vehicles, killed the market for the Long March rocket. The replacement satellite, Optus B3, was launched with an Ariane rocket: all subsequent Optus satellites have also been launched by Ariane.

Telstra, for the first time facing a domestic competitor in the telecommunications field, began actively marketing its international services and moved to offer satellite communications services to Pacific nations. As early as 1990, it provided a regional network for members of the South Pacific Forum, organisation representing the interests of Pacific islands, and continued to expand its presence across the Pacific and in South East Asia, where it had begun offering its expertise in the development of ground stations in the 1980s (see Chapter 8).

The international satellite telecommunications landscape also began to transition to a more competitive environment in the 1990s, with challenges to the international monopolies of INTELSAT and INMARSAT from a growing field of private satellite operators. Under pressure from these competitors, INMARSAT was split in two in 1999, with its operational arm becoming the private company Inmarsat Ltd, while its former international mobile communications regulatory function was transferred to a new intergovernmental body, the International Mobile Satellite Organization (IMSO). INTELSAT, although resisting privatisation for some time, was effectively forced to do so by legislation passed by the US Congress (America being INTELSAT's largest shareholder). In July 2001, after 37 years as an intergovernmental organisation, INTELSAT was privatised as Intelsat Ltd.

Satellite Communications in Australia Today

Following the privatisation of INTELSAT and INMARSAT, Telstra and Optus had to compete in a burgeoning international satellite communications market, which itself has had

BELOW The Australian flag flies on the Ariane 5 payload fairing for the launch of Optus C1 in June 2003. It was the world's largest hybrid communications satellite, serving both civil and Defence communications. (Photo courtesy of ESA/CNES/ARIANESPACE-Service Optique CSG)

to compete with the extensive development of cellular phone networks around the globe and the rapid rise of domestic and international fibre-optic cable networks to support the phenomenal growth of the Internet and digital services. Telstra continues to operate four major satellite ground stations, now known as teleports – at Oxford Falls in Sydney, Bendigo, Victoria, Perth International Telecommunications Centre at Gnangara and at Stanley, Hong Kong – through which it can access over 40 satellites. It also continues to provide TTC&M services to various satellite companies. The company now has a global presence, but with a concentration on the Asia-Pacific and South East Asian regions. Optus continues to offer satellite-based services, in addition to its terrestrial mobile phone and data network, with five satellites currently in orbit serving Australia, New Guinea, New Zealand, South East Asia and Pacific Islands across to Hawaii. Optus' most recent satellite, Optus 10, was launched in 2014.

The two Sky Muster satellites, launched in 2015 and 2016 for NBN Co. Limited, are the latest additions to Australia's satellite-based communications infrastructure. Components of the National Broadband Network (NBN), they provide broadband services to remote areas, as well as Norfolk Island, Christmas Island, Macquarie Island and the Cocos Islands. The first Australian satellites to operate in the Ka band, Sky Muster I and II are designed to provide fast broadband service with speeds of up to 25Mbps. Together, the two satellites service about three percent of the Australian population, around 400 000 Australian homes and businesses, as a means of bridging the 'digital divide' between rural and metropolitan access to broadband services.

BELOW The National Broadband Network's Sky Muster satellites are supported by a network of ground stations across the country. The Wolumla station, south of Bega in New South Wales, is one of two that provide TTC&M services for the satellites. (Reproduced with permission from NBN Co, 2017)

To support the satellites, NBN Co. has established its own network of ground stations in New South Wales (Wolumla, Broken Hill and Bourke), Western Australia (Geraldton, Kalgoorlie, Waroona and Carnarvon - alongside the former OTC site), Queensland (Roma), Tasmania (Geeveston) and in South Australia at the former OTC Ceduna site. The Wolumla and Kalgoorlie stations also have TTC&M capabilities.

The capabilities of the Sky Muster satellites not only bring digital connectivity to rural and remote areas, they also offer the possibility of establishing effective telehealth services in these isolated regions, where regular health and medical services are restricted or unavailable. Although some attempts at providing telehealth services in remote communities were previously undertaken with Aussat and Optus, the limitations of older satellite technologies meant that the service that could be provided was very limited.

International satellite telecommunications have linked the world by invisible threads, facilitating global trade and commerce, providing access to news, entertainment and information content around the clock, and opening formerly isolated regions to a range of public good services, including teleeducation and telehealth. The immediacy of contact that is now available has had a profound effect on the world's cultures and been responsible for unprecedented social and economic change: in some ways bringing to fruition in the Twenty-First Century what philosopher Marshall McLuhan first dubbed 'the global village' in 1968.

The Eye in the Sky: Meteorology and Remote Sensing

Just as the 'high ground' of space offers important advantages for defence and national security surveillance and intelligence gathering, so too does the overview from orbit have a wide range of civil applications. This 'remote sensing', as it is termed, where observations of the Earth are made in wavelengths both visible and invisible to the human eye, can reveal information about our planet, its environment and resources that would not be otherwise apparent. The broad area perspectives available from orbit allow the observation of natural events and phenomena that can only be fully understood when seen and studied at the macro, or even global, level. They also make it possible to survey large areas that might be difficult or prohibitively expensive to study at ground level.

The Forecast for Today: Meteorology satellites

While the weather is always with us, its impact on people's lives extends far beyond the trivial matter of what clothes to wear for the day's conditions: Australians know only too well how flood, drought, bushfire, cyclones and other extreme weather events have both immediate and long-term effects on communities and the economy. Meteorological data is essential information for forecasting and climate services, for defence, aviation and shipping, as well as for the emergency services.

From their vantage points in orbit, meteorology satellites can obtain data on cloud formation, temperature, humidity, wind speed and direction, sea surface temperature and ocean currents, and atmospheric instability over areas previously inaccessible. In addition to routine weather forecasting, satellite observations can locate and track cyclones, assess the likelihood of floods, severe storms and bushfires, and contribute to ongoing meteorological research. Weather satellites are also used to relay data from drifting ocean buoys, atmospheric balloons, and automatic weather stations in remote areas.

The economic and social importance of weather made meteorological observations from space the first satellite remote sensing application. In fact, experimental weather photographs were taken from sounding rockets and balloons as early as 1950, and Chapter 3 has already outlined the sounding rocket programs at Woomera in the 1960s and 70s that made important contributions to the understanding of meteorological conditions in the upper atmosphere above Australia.

The first weather satellite was TIROS 1 (Television and Infra-Red Observation Satellite), launched by NASA in 1960. Australian meteorologists were among the first in the world to receive live weather images broadcast directly from space from the TIROS 8 satellite. Test transmissions were received on Christmas Day 1963, with the first regular transmissions being received on 7 January 1964. This new capability had an immediate and profound impact on the quality of the weather forecasting and warning services provided by the Bureau of Meteorology. For the first time, many weather systems – such as tropical cyclones – were observed that previously would have remained undetected because they developed over ocean areas that could not be observed.

The first satellite photos showing all of Australia in one image were received from NASA's Applications Technology Satellite (ATS) 2 in April 1967. By the late 1960s, real-time satellite imagery was distributed rapidly to Bureau of Meteorology forecasting centres throughout Australia. Satellite reception stations for cloud imagery from Nimbus satellites were operated around the clock in Melbourne, Perth and Darwin, and other stations followed in Brisbane and at Australia's Antarctic bases. Early use of satellite images involved the identification, location, and extent of weather systems, and their development and movement. The ever-improving ability to detect and predict the path of extreme weather events such has cyclones has saved countless lives and helped to minimise damage to property and infrastructure.

Models of satellite-observed low pressure systems were developed to allow objective estimates of meteorological parameters, such as surface pressure, to be made directly from cloud images. Cloud imagery also helped to determine wind speeds and directions: infra-red imagery of clouds provided information on temperatures at different altitudes. During the 1970s, the first accurate quantitative measurements from satellites became

BELOW How weather satellite imagery had changed in 50 years. This 13 February 1965 picture (top) was the first overview of the world's weather, laboriously stitched together from 450 individual images taken by the TIROS 9 weather satellite. Exactly 50 years later NOAA's Suomi NPP satellite produced the image (bottom). Computer tools now routinely fit together the fourteen 3000 kilometre wide swaths of the Earth that are collected each day. (Photos courtesy of NOAA/NASA)

FIRST COMPLETE VIEW OF THE WORLD'S WEATHER

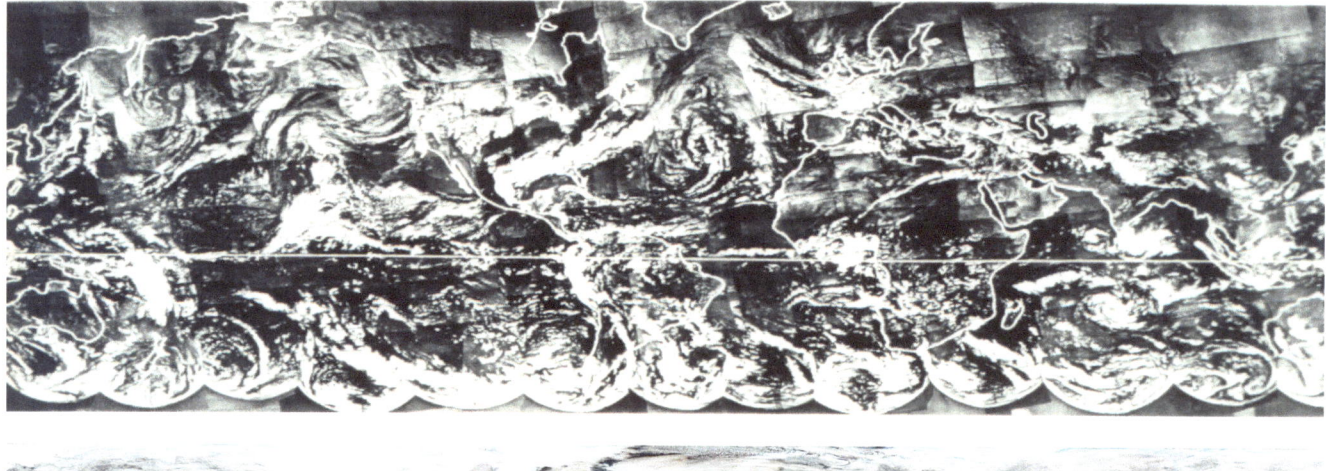

available to Australian meteorologists. This data was processed to provide vertical profiles of atmospheric temperature and humidity, called soundings, and contributed to a steady improvement in the accuracy of numerical weather prediction models in the Australian region. The Bureau of Meteorology opened its direct readout and processing system for sounding data from the US polar orbiting National Oceanic and Atmospheric Administration (NOAA) weather satellites in 1980, receiving data once every twelve hours: today the NOAA satellite data is received every six hours.

Satellite data and computer modelling form a powerful combination in weather prediction. Australia has made extensive use of weather prediction analysis, utilising predictions derived from atmospheric modelling performed by powerful supercomputers, since the 1980s. In 2016, the Bureau of Meteorology commissioned its latest Cray supercomputer, to further improve numerical weather prediction modelling and forecast products. Data from the Himawari-8 satellite (see below) provides significant input into current Australian weather modelling.

Since 1977, a major source of Australia's satellite weather imagery has been satellites of the Geostationary Meteorological Satellite (GMS) Himawari series operated by the Japan Meteorological Agency. Located in geostationary orbit directly north of Australia, these satellites provide a constant view of Australia and the South-East Asian region. They provide both visible and infra-red images: visible images obtainable only in daylight, with 'infra-red' images available 24 hours a day, produced by sensing heat radiation from the Earth and its atmosphere. Initially data from GMS satellites was updated every three hours, which was decreased to one hour by the 1990s.

Above GMS Himawari 8 views of Australia's weather in colour and infra-red taken on June 7, 2017. Data from Himawari 8's multiple wavebands makes an important contribution to the modelling of Australia's weather patterns. (Photos courtesy of Japan Meteorological Agency)

Between 2005 and 2015, Japan's Multi-functional Transport Satellite (MTSAT), a variant of the Himawari series that provided services for air navigation, as well as meteorology, was the major provider of weather imagery to Australia. In 2015 Himawari-8 came into service. This satellite produces about 50 times more data than its immediate predecessor, providing 16 observation wavebands that capture important detail from many layers of the atmosphere. Detailed scans of Australia and nearby regions are delivered every 10 minutes.

A Turn Around Ranging Station (known as TARS 1) was set up in Melbourne at the request of the Japanese Government to assist with 'station keeping' (maintaining an accurate orbital location) for GMS Himawari 1. This type of ground station 'turns around', or receives and retransmits signals sent via the satellite from its main control station. In 1985, the Australian and Japanese governments signed an agreement on the operation of Japan's geostationary meteorological satellites: Australia received satellite data from GMS satellites in return for operating the TARS 2 ground station.

A similar arrangement with China in 1991 allowed Australia to receive data from China's first geostationary meteorological satellite, Feng Yun 2, which provided coverage of Western Australia and the Indian Ocean, in return for hosting a Chinese Turn Around Ranging Station (C-TARS). This led to the establishment, in 1992, of the Bureau of Meteorology's Crib Point Satellite Earth Station, located at HMAS Cerberus Naval Training Base on Westernport Bay in Victoria. Crib Point is the only dedicated weather-related Satellite Earth Station in Australia and exemplifies international co-operation in the meteorological field between Australia, China and Japan. In 1994, Crib Point was upgraded to serve Japan's GMS 4 and GMS 5 satellites (J-TARS). Today, the station supports or receives data from Himawari 8, China's FengYun 2 geostationary satellite series, NOAA polar-orbiting satellites and NASA's Aqua, Terra and Suomi NPP environmental monitoring satellites.

A New World View: Remote Sensing Satellites

Alongside meteorology satellites, remote sensing satellites study the Earth to improve understanding of our planet and its environment, enabling better management of natural resources and the built environment. Remote sensing satellites play a key role in the management and mapping of the vast continent that is Australia.

Remote sensing satellites study the land, sea and air by detecting a wide range of electromagnetic wavelengths (such as infra-red, ultra-violet and radar as well as visible light) reflected from the Earth. Using computer-enhancement techniques, this data can produce images that reveal details and information not visible or discernible to the human eye. Remote sensing satellites provide inestimable quantities of valuable data for such diverse activities as agriculture, mining, environmental monitoring, land-use and urban planning, water resource management, fisheries management, ocean studies, atmospheric research and disaster management.

Below The array of meteorology satellites in orbit in September 2016. Australia's Bureau of Meteorology accesses data from many of them to support weather prediction and environmental research. (Illustration courtesy of NOAA)

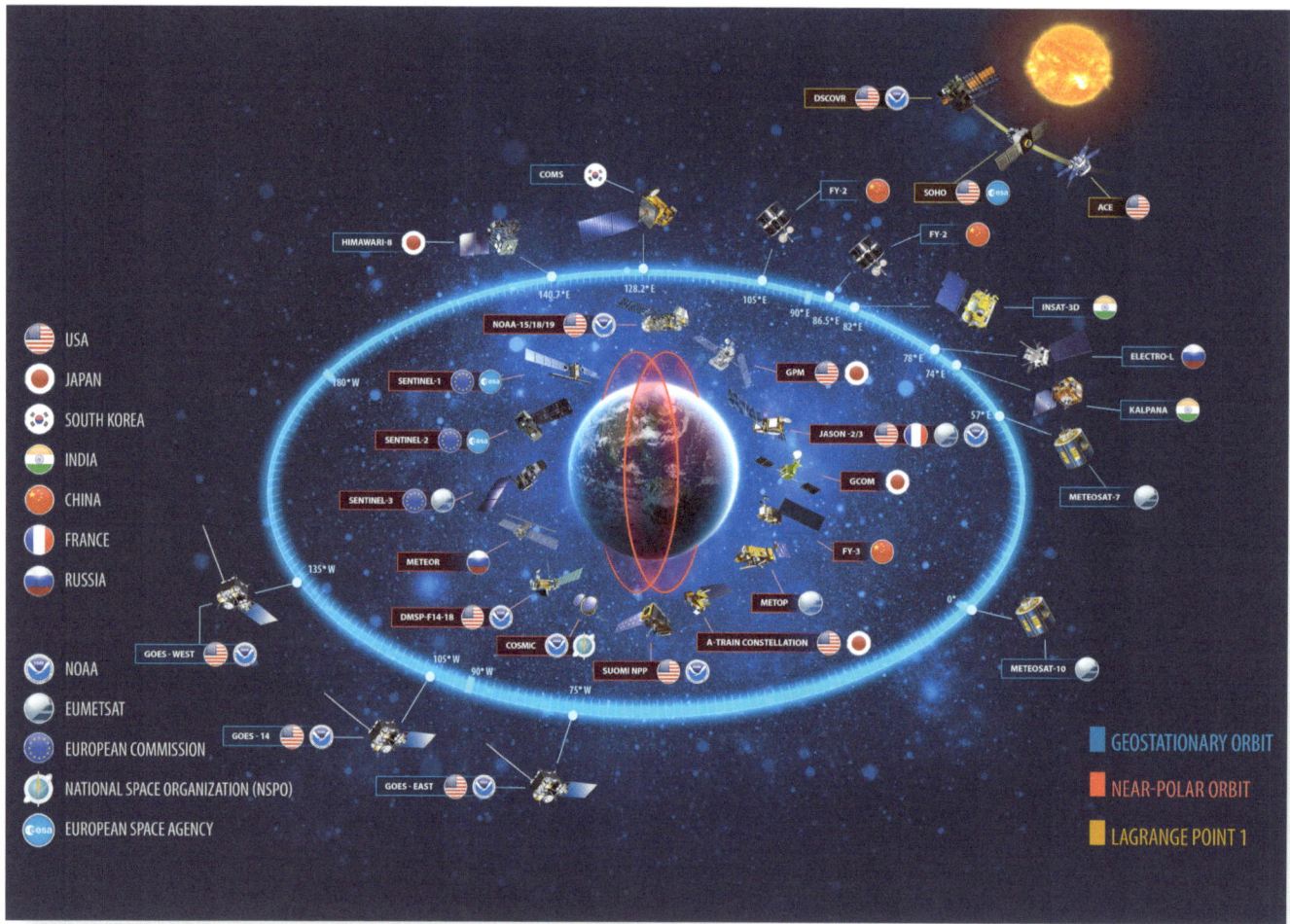

Space Weather

Space weather, a term that became popular in the 1990s, refers to the effects of solar activity on the Earth's upper atmosphere and near-space environment. The Sun sends out a stream of charged particles from its atmosphere, known as the solar wind, as well as high-energy X-rays and high-speed particles resulting from violent eruptions on the Sun's surface. When these reach the Earth, their interaction with the magnetic field and the upper atmosphere produce beautiful aurorae, but they can also have dangerous and damaging effects.

High-energy particles, dangerous to humans, can reach the stratosphere where jet aircraft fly. Space weather disturbances can interrupt HF radio and satellite transmissions; they can damage and even disable satellites, damage power grids and even affect aircraft avionics; they can reduce the life of satellites in Low Earth Orbits by causing the atmosphere to expand, creating friction that slows satellites down, and even potentially damage long-distance pipelines by reducing the efficiency of their anti-corrosion systems. As our reliance on technology increases, the impact of space weather upon orbital and terrestrial infrastructure becomes greater and potentially more dangerous.

In Australia, space weather is monitored by the Bureau of Meteorology's Space Weather Services (SWS) branch. Originally established as the Ionospheric Prediction Service in 1947, SWS manages an extensive network of sensors across Australia, New Zealand and Antarctica to detect space weather conditions. It also operates a solar observatory at Learmonth, in Western Australia, in conjunction with the US Air Force, and the Culgoora solar observatory near Narrabri, New South Wales, which conducts continuous optical and radio observations of the Sun. These Australian observations are combined with data from satellites and observations from other countries to produce advice and warnings on space weather conditions for radio communications, satellite services, airline operations and utilities managers and the Defence, Intelligence and Emergency services.

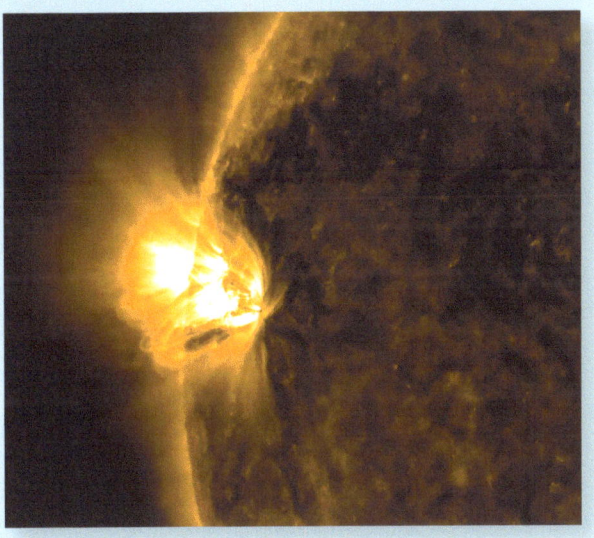

LEFT *Active regions of the Sun like this one observed at the beginning of June 2017 spew out X-rays and high energy particles that can damage satellites and disrupt communications. (Photo courtesy of NASA/SDO and the AIA, EVE, and HMI science teams.)*

Remote sensing satellites generally travel in polar orbits at altitudes of 700-1500 kilometres, although others are located in geostationary orbit and even further away at gravitationally stable positions in deep space. The resolution capabilities (the smallest size of object that can be discerned in an image) of remote sensing satellites have been steadily increasing since the 1970s: where early satellites could only resolve structures around 75m in size, some of today's remote sensing satellites can see objects less than a metre: approaching the capabilities of military surveillance satellites.

While many early weather satellites carried out Earth resources studies, the first satellite specifically designed for remote sensing was ERTS 1 (Earth Resources Technology Satellite 1), also known as Landsat 1, launched by the US in 1972. It marked the beginning of the Landsat series of satellites, which is currently in its eighth generation and now provides almost half a century of data on environmental and land-use change, enabling trends over time to be discerned.

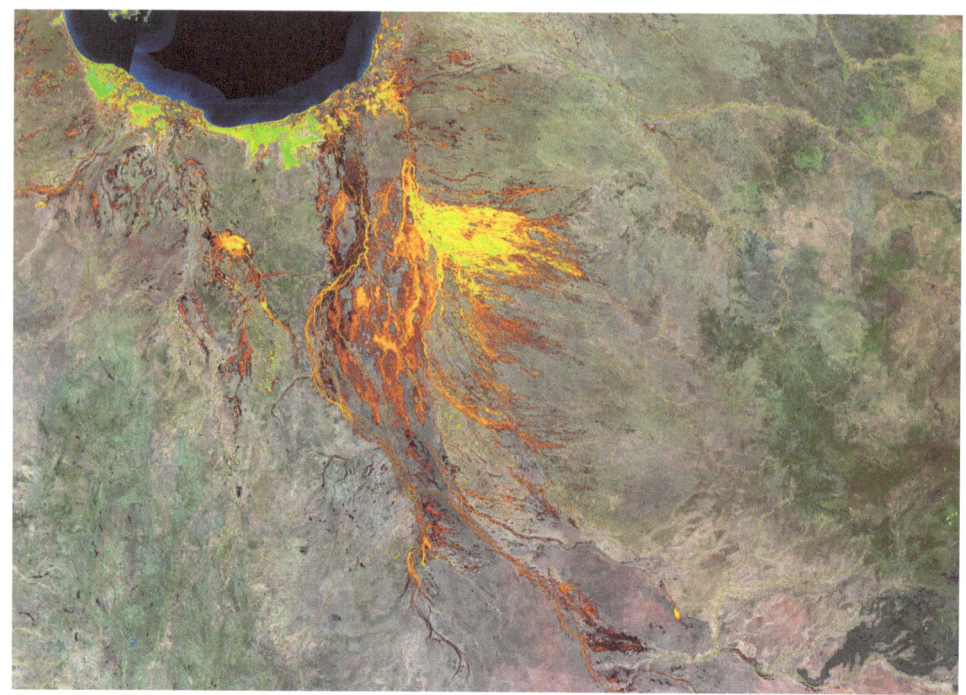

Above Thirty years of data enabled the creation of this image, showing the cumulative inundation of rivers flowing into northern Queensland's Gulf of Carpentaria. False colour is used to highlight the data collected by the satellite using particular wavelengths. (Image © Commonwealth of Australia (Geoscience Australia) 2017)

Recognising the advantages that Landsat could bring to Australia, the Bureau of Mineral Resources, the Bureau of Meteorology and the CSIRO became involved with Landsat from the beginning, funding five years of research using Landsat and other satellite data. By 1975, the CSIRO had obtained the first comprehensive digital archive of Landsat scenes of Australia, and in that same year the Department of Science recommended the establishment of an Australian Landsat receiving station at Alice Springs, with associated processing and archiving facilities at Belconnen, Canberra.

The mining industry quickly recognised the value of remote sensing for mineral exploration and rapidly became not only the largest user of remote sensing data in Australia, but also a strong supporter of remote sensing research and applications development. The strongly voiced support of the industry encouraged the Fraser Government to allocate the funds for the establishment of the Australian Landsat Station (which became operational in November 1980) and a data processing and archiving facility. Renamed the Australian Centre for Remote Sensing (ACRES) in the 1980s, the satellite station was operated by the Australian Surveying and Land Information Group (AUSLIG), which is now part of Geosciences Australia. Over the years the station has been equipped to receive, record and process data from all generations of Landsat satellites: in 2011 the Australian Space Research Program (see Chapter 8) funded a project called Unlocking the Landsat Archive to improve access to this invaluable trove of Australian environmental data, making it available online. The Alice Springs station has been regularly upgraded to receive data from an increasing range of satellites including NOAA satellites, the French SPOT spacecraft, ESA's ERS 1, 2 and Envisat, to which Australia contributed (see Chapter 7) and a wide range of current environmental monitoring satellites. Its 2016 upgrade also added the capability to provide TTC&M services for Landsat satellites.

ACRES is now known as the National Earth Observation Group (NEOG) and continues to operate the upgraded station at Alice Springs and the data processing facility in Canberra, which also processed data from the Tasmanian Earth Resources Satellite Station (TERSS) operated by the University of Tasmania near Hobart. TERSS was a collaborative project, which involved the university and several major Australian organisations in the initial design and building of Australia's second Earth resources satellite ground station. Opened in the 1980s, it operated until 2014, when improvements in satellite technology rendered it obsolete. The Western Australian Satellite Technology and Applications Consortium (WASTAC), comprised of state and federal departments and universities to collect and maintain an accessible archive of environmental satellite data, also began to operate a satellite receiving station at Curtin University in Perth in the 1980s. It now operates a station at Murdoch University that receives data from the Terra,

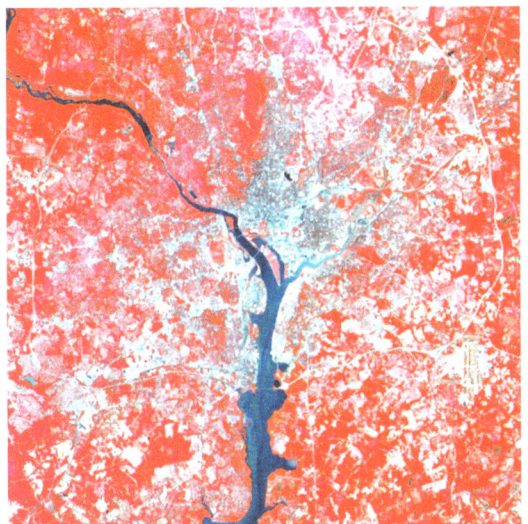 

LEFT *These Landsat images of Washington DC, taken 40 years apart, show the improvement in remote sensing satellite resolution in that time. The 2012 Landsat 7 image shows much greater clarity and detail than the original 1972 view. False colour is used to differentiate between urban areas, vegetation and water. (Image courtesy of US Geological Survey, US Department of the Interior)*

Aqua, Feng Yun 3B, Metop 2, NPP and NOAA satellites. A receiving station for NOAA satellites was established in Townsville in the early 1990s: it was decommissioned in 2014. The Integrated Marine Observing System (IMOS) installed a new reception antenna at the Australian Institute of Marine Science near Townsville in 2008, to extend data coverage into the equatorial part of the Western Pacific Ocean, which are important regions for the El Niño/La Niña phenomenon and tropical cyclone development.

By 1982, the US had developed a new form of Landsat scanner, the 'Thematic Mapper', which required an upgrade to the ACRES station in order to receive the data. However, with the country in recession at the time, the government would not fund the upgrade. Dr Ken McCracken, Chief of the CSIRO Division of Mineral Physics, persuaded the organisation to contribute 25 per cent of the funding and raised the rest of the funds from industry and the research community, ultimately resulting in 24 'shareholders' in the Thematic Mapper project. Along with the first Australian National Space Symposium, held in 1984, the effort involved in the Thematic Mapper project was part of the momentum that led to the development of the CSIRO Office of Space Science and Applications (COSSA), which became actively involved in the development of Australian remote sensing instruments that were flown on satellites in the 1990s (see Chapters 7, 8 and 9 for further aspects of the COSSA story).

The spectral signatures (electromagnetic radiation reflecting properties) of the Australian landscape are unique, so researchers have had to develop new hardware and software to make effective use of the data gathered by spacecraft designed to examine

BELOW *The newest antenna at Geoscience Australia's dedicated remote sensing satellite ground station at Alice Springs. Commissioned in 2016, it is decorated with an artwork called 'Caterpillar Tracks' created by Arrernte artist Roseanne Kemarre Ellis and links the Arrernte people's traditional mapping of their land to the role of the antenna in collecting satellite imagery of the Earth. (Image © Commonwealth of Australia (Geoscience Australia) 2017)*

ABOVE This 2014 Landsat 8 view of Rocklea Dome (the oval in the upper centre) in the Hammersley region of Western Australia reveals iron ore deposits (the meandering features) using the satellite's shortwave infra-red and near-infra-red detectors. (Photo courtesy of the US Geological Survey, US Department of the Interior)

the greener, more temperate regions of Europe and North America. Australian and New Zealand researchers have made great contributions to the field of image enhancement and in the process developed a number of commercial remote sensing software analysis package that are discussed in Chapter 8.

Remote sensing is used in Australia for agriculture, geology, oceanography, surveying, disaster monitoring, land and water management, urban planning, police surveillance (searching for the spectral signatures of concealed marijuana crops, for example), infrastructure development, wildlife studies and environmental and climate change research. Satellite surveillance has proved very valuable to farmers for crop management and has even been used to predict outbreaks of insect pests: applied to pasture management it can identify eroded soils in need of remediation. The mining industry is still one of the largest users of satellite imagery in its search for new mineral deposits and likely oil-bearing strata, while other geologists and geographers have been able to obtain a greatly improved overall picture of the geological make-up of the continent and its geography.

The increasing rapidity with which processed satellite imagery can be made available to the emergency services has become an invaluable tool in bushfire control and flood management. Oceanographers have used satellite data to assist with intensive studies of the Great Barrier Reef and the seas surrounding Australia, while ecologists and a wide range of government, university and private researchers apply satellite remote sensing data to their environmental studies. The vast territory of Antarctica, in which Australia has a special interest, has also been mapped and studied through the use of satellite imagery.

Climate change has become a major focus of remote sensing and environmental research over the past three decades. It was primarily space-based remote sensing systems that began to reveal the global changes occurring as a result of greenhouse gas emissions from industrial activity and energy production using fossil fuels, as well as overpopulation. Space-based environmental surveillance of the Earth continues to increase in importance as the most appropriate way of monitoring global change. NASA's Earth Science

BELOW Landsat 8 image showing bushfires and their aftermath in the Grampians National Park in Victoria in 2014. In this natural colour image, the burned land is grey-brown, while plumes of smoke from active fires can also be seen. (NASA Earth Observatory image by Jesse Allen and Robert Simmon, using Landsat data from the US Geological Survey)

program, originally called Mission to Planet Earth, began in the 1990s as a long-term project of observations from orbit to discover the causes and discern the effects of climate change: other nations have also instituted satellite observation programs concentrating upon the changes that are taking place in it as a result of human activities. Military, Intelligence and economic assessment agencies are also recognising the implications of climate change for defence and national security and its potential effects on the world's economy.

Australian researchers are active in investigating the environment and climate change, utilising the constantly growing range of instruments optimised for this research that have been flown on NASA, Japanese, European and other satellites. Satellite data from NASA's TOMS instrument first identified the 'ozone hole' in the Southern Hemisphere, which allows higher levels of ultraviolet radiation than normal to impact upon Australia in summer, potentially increasing the skin cancer rate in the population. Data from orbit also helped to identify the El Niño/La Niña ocean current phenomenon, which is responsible for both droughts and extreme rainfall events in Australia. As global temperatures rise, Australia is particularly at risk from changing environmental conditions such as rising sea levels and rising temperatures: remote sensing data provides researchers with the keys to understand, predict, prepare for and mitigate where possible the effects of climate change on the nation in the coming decades.

GUIDING STARS: POSITION, NAVIGATION AND TIMING SERVICES

Today, Global Positioning System (GPS) location services are available to everyone and incorporated into smartphones, cameras, cars and an array of other digital devices. The ubiquity of GPS in modern life is such that it is difficult to realise now that it was only twenty years ago that such services became widely available to the public and that initially, precise positioning and navigation services were only available to the militaries of the Cold War Superpowers.

Because the orbits of satellites are precisely known, and they can be seen from most parts of the Earth below, the Doppler effect of a satellite's movement in orbit on a radio signal it sends or receives can be used to determine position on the Earth's surface with much greater accuracy than traditional navigation tools. Navigation systems like GPS use satellites that carry atomic clocks, accurate to one second in 33000 years, and travel

BELOW Measures of sea surface height, which varies with the temperature of the water, helped to identify the El Niño/La Niña ocean current phenomenon. These images from the US-French TOPEX/Poseidon satellite show sea surface temperatures in December 1996 (left) and August 1997 (right). Red and white false colours indicate particularly warm water and the presence of an El Niño event. (Image courtesy of NASA/JPL)

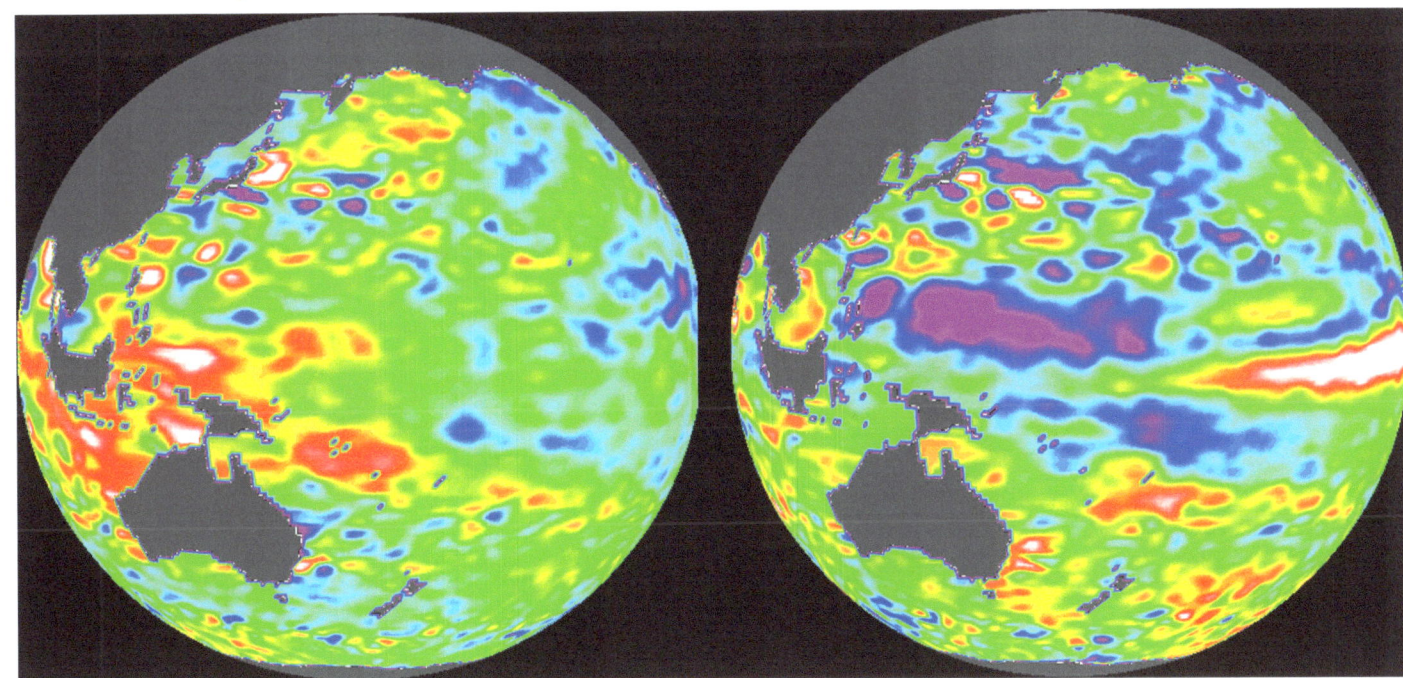

LEFT *Australian researchers discovered that a sensor on ESA's Envisat satellite could detect coral bleaching down to 10m deep. This means that damage to the Great Barrier Reef and other coral reefs under stress can be more rapidly detected and studied. (Photo courtesy of ESA)*

in orbits, which are known with great precision. GPS-enabled instruments also carry (or can access digitally) very precise clocks, synchronised to keep the same time as those carried by the satellites. Consequently, when a ground user receives a signal from a navigation satellite, difference between the transmission and reception times of the signal allow the determination of location in relation to the satellite. Signals from four satellites are needed to accurately locate a position on the ground, although a good estimate of position can be obtained if only three satellites are 'visible' to provide signals.

The first navigation satellites were launched in the early 1960s for military use, to enable naval ships and submarines to navigate with greater accuracy. In conjunction with beacon stations on the ground, these early navigation satellites were also used to better determine the exact shape of the geoid (the roundness of the Earth), in order to improve the targeting of ballistic missiles.

The first US navigation satellite system, Transit, became available for use by maritime shipping in 1967 and was gradually opened to wider civilian use. From the end of the 1980s, the NAVSTAR-GPS system (Navigation System using Timing and Ranging-Global Positioning System) was introduced. The Soviet Union developed a similar system, called Glonass, which has continued to be operated by Russia. Both systems employ networks of satellites to provide global coverage for military and civilian use.

During the Cold War, NAVSTAR-GPS was available to civilian users under a degraded signal system that did not provide positional fixes to the same high level of precision as the military system. Initially, the military system had a precision within five metres, while the civilian GPS signal provided a level of accuracy of around 100 metres. This 'selective availability', as it was known, was discontinued in May 2000, although the US Government retained the capability to re-instate it in the event of major hostilities.

LEFT *The GPS satellite constellation currently consists of 31 satellites, in 20000km orbits that are inclined to the Earth's equator by about 55°. The system is designed to ensure that at least four satellites are visible at least 15° above the horizon at any given time anywhere in the world. (Illustration courtesy of NASA)*

In September 2007, it was decided that future generations of GPS satellites would no longer have the selective availability feature. It was this certainty for the civil use of GPS that enabled the subsequent rapid and widespread growth of GPS applications.

Some Australian warships were fitted with NAVSTAR receivers prior to the end of the Cold War, and civil GPS use began to grow during the 1990s. Before the end of selective availability, geologists, surveyors, oceanographers and environmental researchers used a special technique called High Precision Differential GPS to determine precise locations of geological formations, surveying points and scientific instruments established in remote locations. Australian scientists were among the pioneers in the development of the software for this technique.

While the everyday person might use GPS primarily for wayfinding, geo-tagging locations on social media, or looking for places using Google Earth, the capabilities of the system underlie many aspects of modern life. Global financial transactions, from ATM withdrawals to major transfers of funds between banks, are verified by timestamps provided by the superaccurate time signals used by the

Above Seven of the current 31 GPS satellites in orbit are Block IIRM types, which included a new military signal as well as a separate civil signal as they were designed before the final phasing out of selective availability. (Image courtesy US National Executive Committee for Space-Based Positioning, Navigation and Timing)

GPS system. The police and emergency services use GPS in their fundamental roles of saving lives. Public transport systems, freight and cargo movements, airlines and shipping are all tracked and navigated using GPS or one of the equivalent systems that have been developed by other nations: China has developed its Beidou system; Europe has Galileo; India has its NAVIC, also known as the Indian Regional Navigation Satellite System (IRNSS). Many GPS devices can now access some or all of these systems. The accuracy of mapping and surveying, important in many areas of national development, as well as to Defence and the emergency services, has also been greatly improved by the use of GPS.

The uses of GPS technology are constantly expanding: wildlife researchers use GPS and other satellites to track the local movement and migration patterns of animals and birds, while film producers employ use GPS-enabled cameras to provide precise position fixes for the insertion of computer generated imagery into their latest blockbusters. Precision positioning also allows for the introduction of a wide range of autonomous robotic vehicles: already the University of Sydney is leading the world in the development of robotic agricultural machinery, such as harvesters, and driverless cars may soon be in commercial operation. To say today that 'I'd be lost without GPS' is to understate the world's reliance on this critical satellite-enabled technology.

ALIVE VIA SATELLITE: SEARCH AND RESCUE SYSTEMS

Since 1982, more than 35000 lives have been saved around the world with the aid of satellite-based search and rescue systems. People or craft lost or in distress can be rapidly located by rescuers as long as they are carrying a radio homing beacon that can be detected by dedicated instruments carried on a variety of satellites.

The COSPAS-SARSAT search and rescue system (derived from the Russian for 'space rescuer' and the English 'Search And Rescue Satellite') was developed as an initiative of the USSR, France, Canada and the United States. An outstanding example of Cold War collaboration by the Superpowers, COSPAS-SARSAT commenced operation in 1982, using two Soviet and two American satellites. The system quickly proved its value and in 1988, search and rescue services from space were formalised in the creation of a treaty-based, intergovernmental, humanitarian co-operative, which today is supported by 43 nations and agencies. Detector instruments, carried as additional payloads on a number of satellites, detect and locate emergency beacon signals and forward the information to over 200 countries and territories – all without charge.

From its first four satellites, COSPAS-SARSAT has now expanded into a multi-satellite architecture that comprises more than 45 satellites: there are satellites in polar low Earth orbit, geostationary orbit and medium Earth orbits of around 2000km,

that can detect an activated emergency beacon. There are three main types of emergency beacon: a 406-MHz beacon designed for use in an aircraft, known as an emergency locator transmitter (ELT); a beacon for marine vessels, called an emergency position-indicating radio beacon (EPIRB); and a small beacon that can be carried by an individual, designated a personal locator beacon (PLB).

When a distress beacon is activated in an emergency, one of the search and rescue satellites will receive the signal and relay it to a regional ground station, called a Local User Terminal (LUT). When the LUT receives signals from a satellite, it calculates the location of the distress beacon and then alerts the appropriate rescue services. Australia has been included in COSPAS-SARSAT coverage since 1989, when the LUT for the original LEO network was established at Alice Springs. In 2016, a LUT for receiving signals from the new medium Earth orbit system, that is in operation but not yet complete, was opened at Mingenew, Western Australia: each of its six antennae is dedicated to a different satellite in the system. Rescues initiated in the Australian region are co-ordinated by the Australian Maritime Safety Authority, which houses the 'Mission Control Centre' to which any detected signals from the LUT are passed for action. 283 lives were saved around Australia in 2012, due to COSPAS-SARSAT services.

The orbital infrastructure that satellites provide has become an essential component of the way the modern world functions, so much so that the possibility of damage or destruction of satellite networks by natural space phenomena or military action is a matter of concern at the highest levels. Imagine a day without satellites – it would be very different from today.

LEFT Help is as close as your EPIRB. Australian company GME has been producing compact EPIRB distress beacons for maritime use since COSPAS/SARSAT coverage was extended to Australia. (Photo by D. Dougherty)

Down Under Comes Up Live: Australia Broadcasts to the World

Before the first regular international relay of TV programs to and from Australia began from the OTC Moree Satellite Earth Station in 1968, the earliest satellite broadcasts into and out of the country were considered major events, eagerly watched even when they occurred at inconvenient hours of the early morning.

Down Under Comes Up Live

Shortly after it became operational, a fortuitous circumstance allowed the OTC Carnarvon facility to make history by becoming the ground station responsible for the first satellite broadcast from Australia. Following an agreement between Australia and the UK, it was planned that Australian television stations would be permitted to use the first two days of operation of the first INTELSAT II satellite free of charge. To take advantage of this opportunity, the Australian Broadcasting Commission (ABC) made arrangements with the British Broadcasting Commission (BBC) to record a program at Carnarvon for Australia-wide distribution: in return, the ABC would record interviews with British immigrants for the BBC.

Above A monitor at the Goonhilly satellite station in the UK, showing the image being received from Carnarvon, captioned as being live from Australia. The image was slightly degraded due to the conversion process for British television. (Photo courtesy of Guntis Berzins and www.honeysucklecreek.net)

However, the launch of the first INTELSAT II satellite (Blue Bird), on 26 October 1966, was not successful and the satellite failed to achieve its proper orbit, instead finishing up in an elliptical orbit that limited its commercial use. The initial broadcast idea had to be abandoned, but hasty calculations revealed that, when used in tandem with INTELSAT 1, short segments of television broadcast could be relayed via INTELSAT II. A modified proposal was quickly put together, giving rise to the first live satellite broadcast from Australia, 'Down Under Comes Up Live'.

On the afternoons of 24 and 25 November 1966, Blue Bird became 'visible' to the Satellite Earth Station at Carnarvon and the Satellite Earth Station at Goonhilly Downs in Cornwall, England. Test patterns were transmitted on 24 November with the broadcast itself occurring on 25 November.

TVW Channel 7 Perth sent a pre-recorded news program to English commercial station ITV, but the 'Down Under Comes Up Live' program itself was a direct one-way broadcast (a return signal was not possible for technical reasons) co-produced by the ABC and BBC. As Carnarvon had no domestic television at the time (it did not arrive until 1972), it was not possible to send a television signal from elsewhere in Australia to Carnarvon for transmission to the satellite. Instead, ABC outside broadcast vans and their technical staff

Down Under Comes Up Live: Australia Broadcasts to the World

travelled the 900km from Perth to Carnarvon to produce the program. The vision was sent live to London, but the audio to and from London was transmitted separately by cable via the Pacific and Atlantic Oceans: an extremely complex operation.

Down Under Comes Up Live reunited three families of British immigrants living in Carnarvon, two of whom happened to be employees of the NASA tracking station, with their relatives who had been brought into the BBC studios in London. Commentators Kim Corcoran and Peter Pockley, a noted science journalist, presented a brief introduction to Carnarvon and also interviewed local residents, including the then-Shire President and later Federal Member of Parliament, Wilson Tuckey, a local sheep farmer Clarrie Lewington and 'flying padre' Rev. John McCahon from the Australian Inland Mission. More than twelve minutes of television was broadcast to London.

The transmission began just after 6:25am London time (2:25pm West Australian time) on Friday 25 November 1966 and was seen live in the UK.

However, because there were no television links between Carnarvon and Perth, a copy of the UK broadcast was flown immediately back to Australia, where it was screened a few days later.

Australia Day at Expo 67

Apart from some test transmissions, the first satellite broadcast into Australia was Australia's 'special day' at Expo 67 in Montreal. Several hundred thousand people in Australia watched live through the early hours of 6 June, as the nation took centre stage in Canada, in a ten-hour program that commenced with Prime Minister Harold Holt officially opening the Australian Pavilion. Australian events at the Expo included boomerang throwing, sheep-dog trials, wood chopping contests and tennis matches with members of the Australian Davis Cup team. A variety concert, 'Pop goes Australia', showcased Australian talent, featuring band leader Bobby Limb (who also produced the concert) and performances by Normie Rowe, Rolf Harris and The Seekers.

BELOW *Perth TV station TVW Channel 7's small outside broadcast van with the Carnarvon 'sugar scoop' antenna on the day of the 'Down Under Comes Up Live' broadcast. (Photo courtesy of Guntis Berzins and www.honeysucklecreek.net)*

The first direct telecast across the Pacific from North America, the Expo broadcast surprised Australian audiences with the clarity of the picture: hundreds of viewers rang a Sydney television station wanting to be assured that the vision really was being broadcast live from Canada!

Our World

Just three weeks after the Expo broadcast, on 26 June, Australia participated in the Our World program, the first live global television broadcast, which linked 24 countries via four communications satellites. Intended as a major achievement for both space technology and international relations – as the USSR and several Eastern European nations were originally going to participate, before pulling out of the broadcast as a protest against the Six-Day War in the Middle East – Our World was a potent demonstration of the potential reach of satellite television, with a worldwide audience estimated between 350 and 700 million. As part of the United Kingdom contribution, The Beatles, at the height of their global fame, gave the first performance of '*All You Need Is Love*' as the finale to the broadcast.

Australia was one of 14 countries to present material during the broadcast. Since it took place between 5:00am and 7:00am Australia's contribution commenced at 5:22am local time, with a visit to the Hammer Street Tram Depot in Melbourne, where the first tram of the day was departing. This segment was the first cross to the Southern Hemisphere and came directly after a broadcast from Japan. At the time, Our World was the most complex television transmission ever attempted, with the switch from Japan to Australia being the most technically complicated of the program, as the Japanese and Australian satellite stations had to switch immediately from transmission to receiving mode and back again.

Later Australian segments presented during the program were a tour of the CSIRO's 'Phytotron' plant laboratory, and radio astronomy observations with the Parkes Radio Telescope. The broadcast was beamed into and out of Australia via NASA's Cooby Creek facility, and transmitted around the country with assistance from OTC, the PMG and the Department of Supply.

BELOW An ABC cameraman, wearing a heavy jacket against the early morning winter cold, prepares for filming the Our World segment at the Parkes Radio Telescope. (Photo courtesy of CSIRO)

ABOVE NASA astronaut Andy Thomas, the first Australian citizen to make a spaceflight, holding a copy of the Sydney 2000 Olympic torch, which he arranged to have flown on the Space Shuttle. (Photo courtesy of NASA)

BELOW An image taken by the Australian-designed Large Format Photon Counting Detector, which was proposed for three different space missions. It shows the 30 Doradus nebula in the Large Magellanic Cloud taken in hydrogen light. (Image courtesy of Mt Stromlo Observatory, ANU Research School of Astronomy and Astrophysics)

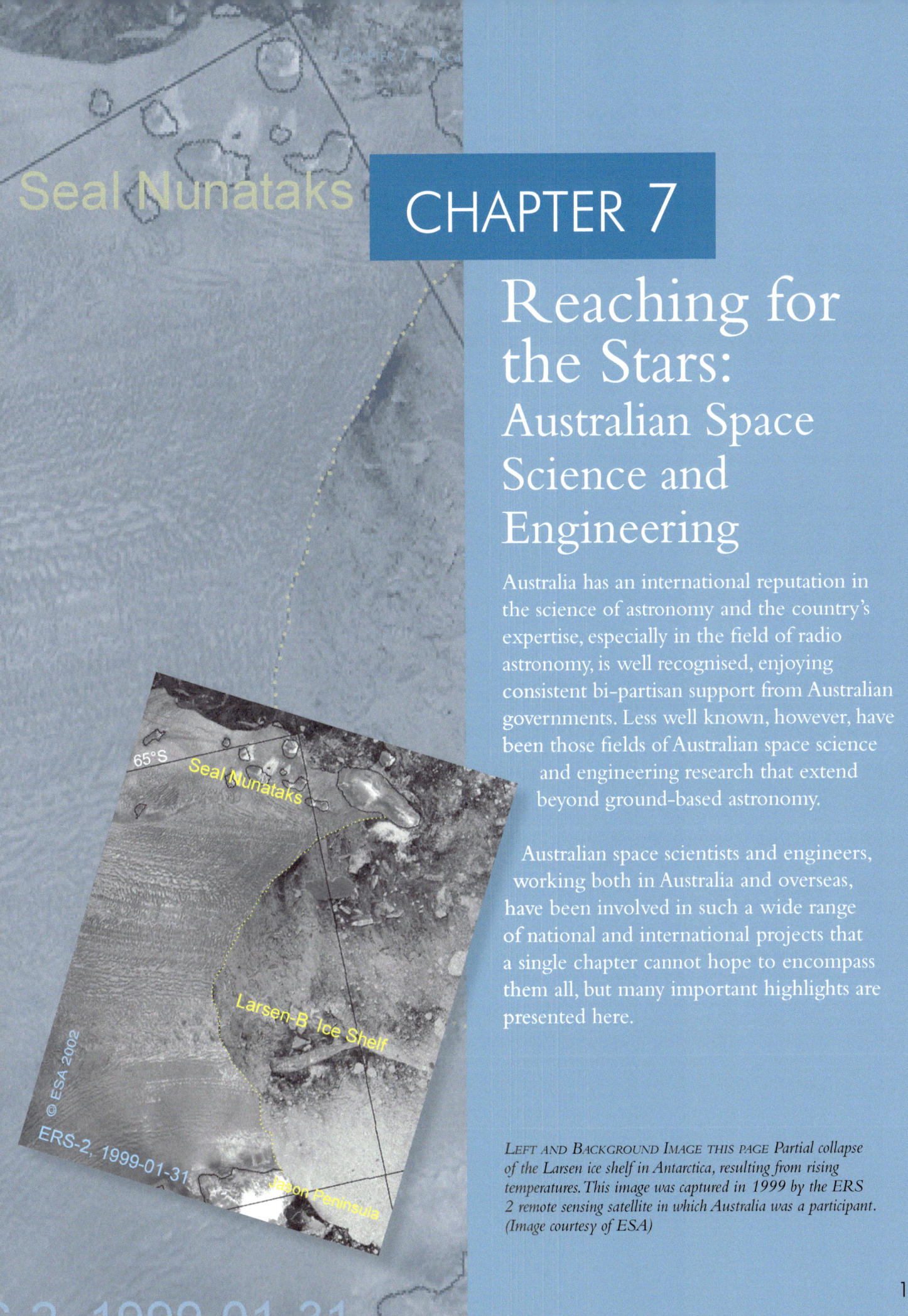

CHAPTER 7

Reaching for the Stars: Australian Space Science and Engineering

Australia has an international reputation in the science of astronomy and the country's expertise, especially in the field of radio astronomy, is well recognised, enjoying consistent bi-partisan support from Australian governments. Less well known, however, have been those fields of Australian space science and engineering research that extend beyond ground-based astronomy.

Australian space scientists and engineers, working both in Australia and overseas, have been involved in such a wide range of national and international projects that a single chapter cannot hope to encompass them all, but many important highlights are presented here.

LEFT AND BACKGROUND IMAGE THIS PAGE *Partial collapse of the Larsen ice shelf in Antarctica, resulting from rising temperatures. This image was captured in 1999 by the ERS 2 remote sensing satellite in which Australia was a participant. (Image courtesy of ESA)*

An important aspect of the Australian space science and engineering story has been the collaboration between the disciplines, particularly since the 1980s, in efforts to establish an Australian space industry (further discussed in Chapter 8) and to engage successive Australian Governments in adopting a national civil space policy and establishing a formal space agency. As Chapter 9 outlines, the fluctuating Australian Government engagement with space activities has, in turn, had an impact on funding for space science, engineering and industry development.

The Beginnings of Australian Space Science and Engineering

Even before the International Geophysical Year, space-related physics research was conducted in the universities and through the CSIROs Radiophysics Division and Meteorological Physics Research Section. The IGY itself, with its focus on the study of Earth's relationship to the space environment surrounding it, provided strong impetus toward the development of space science research in Australia, particularly encouraged by Dr David Forbes Martyn, who, as already noted in Chapter 2, was instrumental in organising Australian participation in the global science program.

In 1958, Martyn became Chief of the CSIRO's Upper Atmosphere Section, a research centre established specifically for him, which enabled him to pursue his long-time interest in the physics of the ionosphere. Martyn would lead this Section until his death in 1970. Between 1963 and 1966, CSIRO flew geophysical research instruments on sounding rockets in the US. Another American sounding rocket campaign was undertaken in 1970 to measure Very Low Frequency (VLF) radio noise in the ionosphere. Although the Upper Atmosphere Section would be closed following Martyn's death, CSIRO interest in space science continued, undertaken within several of the organisation's divisions.

As earlier chapters have described, the IGY gave rise to the British and Australian sounding rocket programs at Woomera, which enabled researchers at the Universities of Adelaide and Tasmania to undertake significant upper atmosphere and space astronomy research programs. The Australian sounding rocket program, together with the WRESAT project, also provided the opportunity for WRE engineers and scientists to apply to space research the technical skills that had been developed on weapons projects, becoming Australia's first space engineers.

When the Australian sounding rocket program was refocused toward specific defence research, afterced the Interdepartmental Committee review of 1970, purely civilian space science research was deprived of the 'free ride' it had previously enjoyed and was required to pay for its rocket flights. Although Professor Carver obtained some grant funding toward launch costs, the fees quickly consumed these funds and the University of Adelaide program of upper atmosphere physics ceased even before the termination of the WRE sounding rocket

BELOW Skylark rockets launched at Woomera carried space astronomy payloads for the Universities of Adelaide and Tasmania in the 1960s and 70s. (Photo courtesy of DST Group)

program. It was, however, asked to supply ozone sensors for NASA's international ozone-sonde sounding rocket campaign in 1978 and 1979. The University of Adelaide space astronomy program, which launched on Skylark rockets, also provided instruments to measure electrical conditions in space in the vicinity of the Sun on the two NASA-West German Helios spacecraft, launched in 1974 and 1976.

BELOW Two identical NASA-West German Helios solar probes, launched in 1974 and 76, carried instruments provided by the University of Adelaide's space astronomy program. (Courtesy of DLR-Archiv Köln)

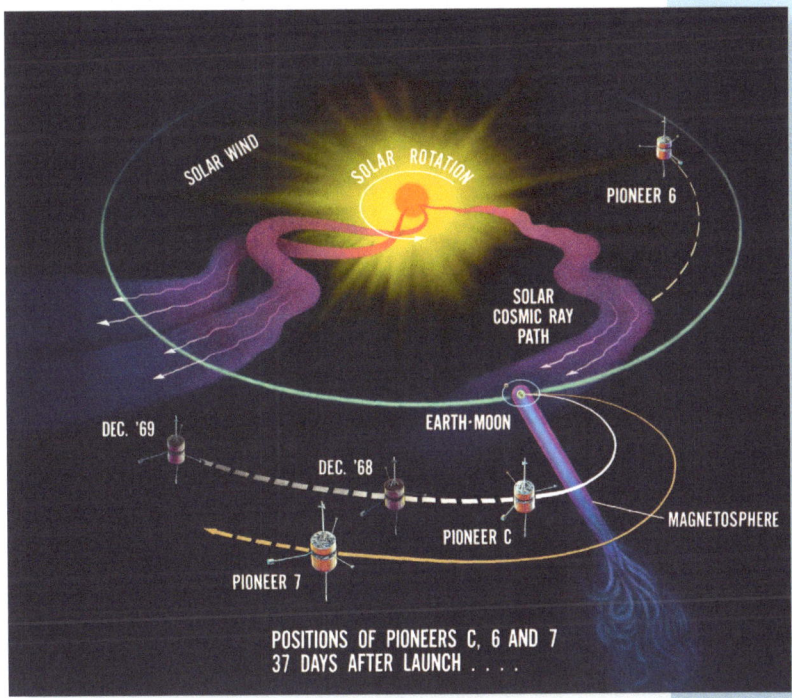

ABOVE NASA illustration depicting the locations in interplanetary space of the Pioneer 6, 7 and 8 (Pioneer C) spacecraft 37 days after launch. Each of these probes carried cosmic ray detectors designed by Australian physicist Ken McCracken. (Illustration courtesy of NASA)

'SIR LAUNCHALOT'

One of Australia's most distinguished space scientists, Dr Kenneth G McCracken earned himself the nickname of 'Sir Launchalot' in the 1960s due to the number of space projects on which he was able to place his research instruments. Responsible for the development of a neutron monitor network in Australia for the International Geophysical Year, McCracken undertook Postdoctoral research at the Massachusetts Institute of Technology (MIT) and was a consultant to NASA's Goddard Space Flight Centre in 1961. His work there, on the hazards of space radiation to astronauts, earned him a 'cameo' in the James Michener novel, Space (1982), in the guise of a young scientist who brings the issue of space radiation to the attention of NASA's Apollo lunar program planners.

As a Professor at the University of Texas (1962-72), McCracken was selected as Principal Investigator for NASA's Pioneer 6, 7, 8, and 9, and IMP D and E spacecraft (launched between 1965 and 1968), for which he designed and built instruments to measure certain characteristics of cosmic-rays. He also built and flew a balloon-borne X-ray detector from India in 1964, making an early discovery of a space X-ray source. Between 1966 and 1969, he also held a Professorship at the University of Adelaide, during which he was able to secure launches on Skylark rockets at Woomera for both Adelaide and the University of Tasmania. This research led to the discovery that X-ray objects are highly time variable. McCracken returned to Australia in 1972 to take up a position with the CSIRO, becoming a significant figure in the development and utilisation of remote sensing in Australia and the founding Director of the CSIRO Office of Space Science and Applications. In 1995, Dr McCracken's role in CSIRO's contributions to the field of remote sensing was recognised with the award of the Australia Prize, together with Dr Andrew Green and Dr Jonathan Huntington.

Up, Up and Away: Balloon-borne Research

For Australian universities without access to the sounding rocket programs at Woomera, stratospheric ballooning offered another way to carry out space science research. The University of Melbourne had commenced a program of upper atmosphere research using high-altitude balloons prior to the IGY, while the University of Tasmania commenced its own scientific ballooning program in 1959, continuing and extending its IGY program of ground-based cosmic-ray research (which had produced one of the world's largest cosmic-ray detector networks).

In 1960, the Australian Balloon Launch Station (ABLS) was established by the Department of Supply at Mildura, Victoria, to support the US HIBAL (High Altitude Balloon) project. This was a program to sample stratospheric air and dust, studying the levels of radioactivity in the stratosphere resulting from US nuclear tests in the Pacific Ocean. Over 600 balloon flights took place between 1960 and 1966, occasionally piggybacking small scientific instrument packages from universities and the CSIRO. The University of Tasmania also conducted a balloon campaign there in 1964-65, before taking advantage of Skylark launch opportunities between 1967 and 1970.

When Project HIBAL was extended, the new Australia-US agreement allowed for balloon flights at other sites besides Mildura and also permitted other agencies to use the launch station in support of basic research into atmospheric phenomena and space astronomy. Overseas researchers were eager to take advantage of the availability of balloon sites in Australia, which offered good observations of astrophysical sources not easily visible (or not visible), from the Northern Hemisphere. Almost 150 balloon flights were carried out between 1967 and 1980 from Mildura and other sites including Alice Springs, Longreach and Broken Hill, carrying gamma ray, X-ray and infra-red astronomy, and cosmic-ray physics payloads. In 1976, the University of Tasmania and Imperial College London constructed one of the largest X-ray astronomy payloads up to that time, which was flown from Mildura (1976), Alice Springs (1978, 1980) and Brazil (1983). A successor project, with collaboration from the Australian Defence Force Academy (ADFA), the University of Tubingen (Germany) and the Italian Space Agency, was launched from Alice Springs in 1986.

In 1980, ABLS Mildura was closed and balloon launching operations were shifted to ABLS Alice Springs, originally established as an outstation of Mildura in 1974. The very low population density over Central Australia allowed flights of larger balloons and heavier payloads than Mildura. It was also better situated for observing the galactic centre and had seasonal wind-patterns that enabled very long duration flights. Responsibility for operating the station was now transferred to the Physics Department at the University of Melbourne. In 1986, the University of Melbourne balloon team transferred to ADFA in Canberra, which has continued to manage ABLS Alice Springs into the present.

Above NASA's Nuclear Compton Telescope gamma-ray research instrument being readied for launch by stratospheric balloon at ABLS Alice Springs in 2010. (Photo courtesy of Ravi Sood, UNSW ADFA)

In 1981, NASA's scientific ballooning program commenced a long-term association with ABLS Alice Springs, resulting in two programs being conducted in parallel: Australia launched its own payloads, as well as those of international collaborators, while the NASA team launched mainly US payloads. Dozens of flights focused on infra-red, X-ray and gamma ray astronomy, as well as measurements of cosmic-rays and other high energy particles. The discovery of the SN1987A supernova in the Large Magellanic Cloud in 1987 brought a sharp increase in balloon missions in Australia, with international teams eager to make observations that could only be performed in the Southern Hemisphere. NASA also used ABLS Alice Springs for test flights of its Ultra-Long Duration Balloon (ULDB) in 2001 and 2003.

The facilities at the Alice Springs balloon station were extensively upgraded between 2008 and 2009 and recent balloon-borne research has included X-ray and gamma ray studies of pulsars, black holes and other exotic celestial bodies. NASA's most recent balloon exercise at Alice Springs was the successful flight and recovery of the HERO (High Energy Replicated Optics) X-ray telescope in 2011. In 2015, a team from Japan carried out a gamma ray experiment, while 2017 saw the return of CNES, the French space agency, to Alice Springs with a campaign of three balloon launches. Further Australian and international balloon launches are planned for ABLS Alice Springs in coming years.

SUPPORTING APOLLO: NOT JUST SPACE TRACKING

Alongside the crucial role played by the NASA tracking stations in Australia in supporting the Apollo lunar landing program, and the many former WRE staff and other Australian scientists and engineers who found their way into positions at NASA during the 1960s, a handful of Australian scientists made special contributions to the Apollo program that are worthy of individual highlight.

Dr John Colvin, Ophthalmologist

A qualified pilot and consultant ophthalmologist to the Royal Australian Air Force, Colvin had designed spectacles that allowed far-sighted pilots to read instruments, enabling them to continue to meet the visual standards required of military pilots. His polycarbonate-lens spectacles, that could withstand high gravitational forces, were adopted by NASA for astronaut use. To shield the Apollo astronauts' eyes from the harsh glare of the Sun in interplanetary space, Colvin designed wrap-around, anti-glare sunglasses as well as 'instrument hood' glasses that were designed to be worn during rendezvous and docking operations where clear forward vision was essential: they limited vision to forward gaze only and reduced interference from reflected light within the spacecraft. Colvin also designed an easily visible docking target for use on the Apollo 7 mission, to assist with practising space rendezvous and docking techniques prior to the Moon landings.

Above Docking target designed by Australian ophthalmologist Dr John Colvin used for rendezvous and docking practice during the Apollo 7 mission. The target was mounted on the S IVB stage of the Saturn IB launch vehicle. (Photo courtesy of NASA)

Dr Brian O'Brien, Physicist

With a Doctorate in experimental cosmic-ray physics, O'Brien took up a position as Associate Professor of Physics at the State University of Iowa, where, between 1959 and 1963, he was able to fly experiments on the Explorer 7, 12 and 14 satellites. He also led small student teams building the Injun 1, 2 and 3 satellites, intended to observe radiation and magnetic phenomena in the ionosphere. Between 1963 and 1968, as Professor of Space Science at Rice University Houston, Texas, he developed several sounding rocket payloads and a small satellite, Aurora 1.

O'Brien began lecturing astronauts on radiation hazards in 1964 and became the Principal Investigator for the Charged Particle Lunar Environment Experiment (CPLEE) that would fly on Apollo 13 and 14. O'Brien believed that lunar dust would be a major hazard to astronauts and, in 1966, developed a matchbox-sized dust detector, which was carried on Apollo 11, 12, 13, 14 and 15. Despite the loss of the experiments on Apollo 13, he had five instruments deployed on the lunar surface transmitting data back to Earth between 1969 and 1976. O'Brien's experiments revealed the hazard that the extremely fine lunar dust will represent to future lunar operations. The tiny grains cling to surfaces, penetrating space suits and causing batteries to overheat; they are also toxic if inhaled. In all, O'Brien's research revealed 14 new findings about the characteristics and behaviour of lunar dust, including that its stickiness changes during the lunar day, that remain relevant today for all future activities on the Moon.

Professor Ross Taylor, Geochemist

New Zealand-born Taylor, a research geochemist at ANU, carried out the first analysis of lunar samples returned to Earth by the crew of Apollo 11. Although Taylor was not originally a member of the ream of scientists assembled by NASA to analyse the samples that would be returned from the Moon, chance led him to assume a leading role in the program. On a visit to the newly-established Lunar Receiving Lab at NASA's Johnson Space Centre in Houston, Taylor demonstrated that he was more knowledgeable about geochemical analysis than any of the selected team and was immediately asked to join the Preliminary Examination Team for Apollo 11. He became manager of the Lunar Receiving Lab and carried out the first spectrographic analysis of the returned lunar samples.

When the first Apollo 11 samples arrived in Houston, Taylor had only a little more than four hours to complete a preliminary analysis of the rocks, as a result was required for a major press conference. Often working 20-hour days over the following weeks, Taylor undertook more detailed analyses, presenting results at a press conference each afternoon. After returning to Australia, Taylor remained a NASA Principal Investigator for lunar samples from 1970 until 1990, working on models for lunar composition and the evolution and origin of the Moon, with access to original Moon rock samples for analysis.

Left Dr O'Brien's two experiments on the lunar surface during the Apollo 14 mission. The Charged Particle Lunar Environment Experiment is in the foreground. O'Brien's tiny dust detector is the small white box on the upper right hand corner of the Apollo Lunar Surface Experiments Package central station in the background. (Photo courtesy of NASA)

Professor John Lovering, Geologist

NASA distributed Apollo mission lunar samples to researchers around the world for study, and Australian geologist John Lovering, at the time Professor of Geology at the University of Melbourne, also received Apollo 11 and 12 Moon rock samples. His chemical analysis revealed a new mineral, Tranquillityite, which was subsequently found in samples from every Apollo mission, as well as a lunar meteorite. Initially thought to be unique to the Moon, Tranquillityite has now been found on Earth as well – at six localities in the Pilbara region of Western Australia.

STARLAB: A RENAISSANCE FOR AUSTRALIAN SPACE SCIENCE AND INDUSTRY

The demise of the sounding rocket programs at Woomera considerably curtailed space science launch opportunities outside the scientific ballooning program, as government science and technology policy made funding difficult to obtain during the 1970s. However, a significant revival in Australian space science and engineering began in 1980 with government support for Australian participation in the Starlab ultraviolet wide-field astronomy satellite project.

Starlab began as a US-Canada proposal for a 1-metre diameter orbital telescope intended to provide very high resolution ultraviolet imagery and spectroscopic analysis that would pinpoint areas that would be recommended for closer examination by more powerful instruments, such as the Hubble Space Telescope. Australia's involvement arose because the Australian National University's (ANU), Mount Stromlo Observatory had considerable expertise in the development of photon-counting detectors, the type of instrument needed for the satellite telescope. It had already developed the Large Format Photon Counting Detector, which was more advanced than any other detector of its type in the world. Canada and the US saw this instrument as crucial to the success of the Starlab concept.

In 1980, ANU convened the Australian Space Industry Symposium, to assess the level of space capability in Australian industry that might support the Starlab program. More than 100 companies expressed interest in the project: when the Australian Government declined in 1981 to fund the design studies for Australian participation in Starlab, fifteen local companies supported it with financial backing for the research and development that was needed.

However, Starlab's potential to provide space qualification experience to local industry, which Australian companies had lost as the Joint Project wound down, eventually drew Government support for the project. It believed that experience acquired on the Starlab project would bolster the opportunities for Australian industry to participate over the long-term in the Aussat satellite communications system. Staff from the Mount Stromlo Observatory worked on Starlab between 1980 and 1984, eventually splitting away from ANU to form the Australian space company Auspace Pty Ltd.

Disappointingly, just as Starlab was building a new momentum for space science, engineering and industry in Australia, the project was cancelled when Canada withdrew its funding due to a switch in that country's space policy toward support for the Space Station Freedom project (which ultimately became the International Space Station). However, as other stories in this chapter and in Chapter 8 will show, Starlab's legacy continued across the 1980s, encouraging a renewed Australian engagement in space activities.

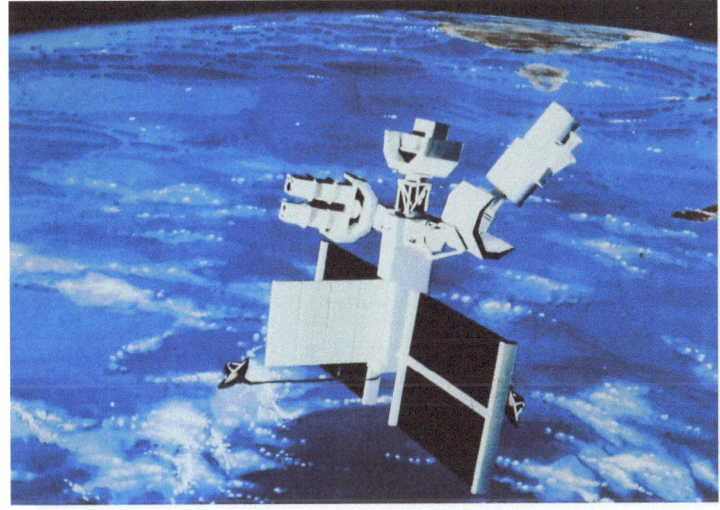

ABOVE Illustration of the proposed Starlab ultraviolet space telescope. Australian involvement in Starlab was responsible for a revival of local space science and industry during the 1980s. (Illustration courtesy of Mount Stromlo Observatory, ANU Research School of Astronomy and Astrophysics)

Fortunately, the developmental work undertaken for Starlab was not lost. Following the project's demise, Australia prepared to participate instead in the NASA/European Space Agency proposed FUSE/Lyman ultraviolet telescope, intended to investigate the origins of the universe by studying gaseous regions and radiation in galaxies. For this project, Australia could provide a near-identical instrument to that developed for Starlab. A one-year feasibility study of the mission was carried out by Auspace and ANU, funded by the Australian Space Office (ASO). The ASO had been created in 1987 to encourage the development of an Australian space industry by funding and facilitating projects that would enhance national space capabilities (see Chapters 8 and 9). The feasibility study commenced in May 1987, with the FUSE/Lyman project office and staff transferred from ANU to Auspace. Once again, however, disappointment was to follow, after the ASO declined in 1988 to fund any further involvement in the project, due to its limited budget.

The Endeavour Space Telescope: Australian Space Science Returns to Orbit

Although Starlab and FUSE/Lyman did not proceed, the research and development for these projects lived on as the basis of the Endeavour space telescope, which would finally return Australian space science to orbit, 25 years after WRESAT. The Endeavour space telescope was originally intended to be the space-qualifying version of the proposed ultraviolet detector hardware for the FUSE/Lyman project. However, after Australia withdrew from that project, Endeavour became a space industry qualification project in its own right, funded by the Australian Space Office within its National Space Program budget.

The Endeavour telescope consisted of a small but complex 100mm binocular ultraviolet telescope housing a Large Format Photon Counting Detector, with associated controls and data recorders. Developed entirely in Australia, principally by Auspace and the ANU, Endeavour was designed to be housed in two Space Shuttle 'Get Away Special' canisters (GASCANs). These containers, about the size of a household garbage bin, were used to carry small, self-contained experiments in the cargo bay of the US Space Shuttle. They were used by many institutions as a way of obtaining cheap access to space for scientific experiments. For Endeavour, one GASCAN contained the telescope and its control electronics, while the second contained the data video recorders and power supply.

Although originally proposed for launch on the Space Shuttle in 1988, as an event for the Australian Bicentenary, the loss of the Space Shuttle Challenger in 1986, and the subsequent delays in the US space program, pushed the flight back to January 1992, when Endeavour was included on the STS 42 mission, aboard the Shuttle Discovery. The Endeavour flight had three main aims: to obtain ultraviolet images of violent events in nearby galaxies, as a contribution to Australian astronomical research; to allow Australian engineers and industry to develop the skills necessary to design and manufacture space systems and hardware; and to space qualify the ultraviolet instrument for use in larger space telescopes.

The Endeavour flight, however, proved to be controversial. The Australian Space Office had accepted the launch opportunity aware that the telescope could not be pointed at individual stars on the flight, due to other, higher priority, requirements of the Space Shuttle. This meant that no astronomical images could be obtained and thus the flight could not fulfil one of its main aims. In addition, the detector only operated for two of the four scheduled observations, as its safety systems detected temperatures in the cargo bay higher than the operational limits of the telescope and so would not permit it to operate. Although critics panned the mission because of these apparent 'failures', the fact that the safety systems operated as they should have done proved that Endeavour was capable of functioning correctly while in space and probably saved it from self-destructing. Consequently, the detector technology was able to be space-qualified, even though it could not achieve its scientific objectives.

Due to the limitations of Endeavour's original flight, NASA offered a second flight opportunity on the STS 67 mission. Since the primary payload was the Astro-2 Spacelab, dedicated to ultraviolet astronomical observations, this would allow the telescope to achieve its scientific objectives. Aboard the Space Shuttle Endeavour, named – like the space telescope – for the ship in which the British navigator Captain James Cook had explored the east coast of Australia in 1770, the Australian space telescope returned to orbit in March 1995 and this time achieved all its original project goals. Despite the success of the mission, the Large Format Photon Counting Detector it carried was no longer at the cutting edge of technology and there was doubt that it could now become a profitable space industry product. The project was consequently terminated, along with the National Space Program and the Australian Space Office, by the Howard Government in 1996.

Left The Endeavour space telescope flies for the second time on the STS 67 mission in 1995. The two GASCANs containing the telescope and its data recorders can be seen attached to the left-hand side of the Shuttle Endeavour's cargo bay. (Photo courtesy of NASA)

CSIRO Office of Space Science and Applications: Promoting Remote Sensing from Space

The groundswell of interest in space science and the development of an Australian space industry generated by the Starlab project coincided with the campaign mounted by Dr Ken McCracken from CSIRO to raise funding from industry and the research community for an important upgrade to the ACRES satellite station, as described in Chapter 6. Catalysed by the level of interest from the scientific, engineering and industry sectors demonstrated by the presence of around 200 local and international delegates at the National Space Symposium convened by the Commonwealth Department of Science and Technology in March 1984, CSIRO decided to focus and co-ordinate its space science and remote sensing activities by establishing the CSIRO Office of Space Science and Applications (COSSA). Its aim was to maximise the environmental, social and economic benefits to Australia arising from research in space-related science and engineering. With a modest annual budget in the vicinity of $6 million, COSSA carried out research in remote sensing systems, communications, radio astronomy and other related fields, with links to many international space programs.

Under its founding Director, Dr Ken McCracken, COSSA very quickly became involved in the development of remote sensing instruments for satellites, with the eventual goal of producing instruments optimised for the unique spectral signatures of the Australian environment, flown on Australian satellites in orbits selected to meet Australia's specific national needs, rather than relying wholly on data provided by the satellites of other nations.

An opportunity for COSSA to participate in the development of a remote sensing instrument was immediately available. The Along Track Scanning Radiometer (ATSR), itself designed by an Australian, was being developed by the UK for inclusion in the European Space Agency's ERS 1 radar remote sensing satellite. Short on funds to complete the instrument, the British approached McCracken to contribute to the ATSR as a way of funding its completion. The Along Track Scanning

Radiometer was a multi-channel imaging radiometer: its main purpose was to provide data concerning global sea surface temperature and to monitor and carry out research into the behaviour of the Earth's climate.

When the National Space Program was established in 1986, as a result of the 1985 Madigan Report (see Chapter 9), the Australian contribution to ATSR 1 was one of the first projects it funded. In addition to CSIRO involvement in the design of the ATSR 1 instrument, COSSA and British Aerospace Australia designed and built its Digital Electronics Unit, which processed data from the ATSR and added it to the data from other instruments on board the satellite. CSIRO was also involved in calibrating and developing software for the instrument. ERS 1 was launched in July 1991 and operated until March 2000, far exceeding its expected lifespan.

The success of the COSSA collaboration on ATSR 1 led to Australia assuming an increased role in its successor instrument, ATSR 2, for the ERS 2 satellite. Auspace was contracted to undertake the partial design and fabrication of the Focal Plane Assembly for ATSR 2, while British Aerospace Australia developed electrical ground support equipment for the testing and calibration of the instrument prior to launch. ERS 2 was launched in April 1995 and operated until September 2011.

Although an opportunity to participate in the development of an instrument for the SPOT 4 satellite was declined, before the ASO was terminated in 1996 it had agreed to support a significant Australian contribution, to the level of co-ownership of the instrument with the UK, to the Advanced Along Track Scanning Radiometer (AATSR). The AATSR was to be flown on board the European Space Agency's Envisat, the successor

BELOW ERS 1 and 2 were near identical ESA radar imaging satellites. After its successful involvement in ERS 1, Australia increased its participation in the ATSR 2 instrument on ERS 2. (Illustration courtesy of ESA)

to ERS 1 and 2, in order to provide continuity of the ERS research data. Auspace designed and constructed the Scan Mirror and associated control units, which were part of the AATSR's Focal Plane Array. Envisat was launched in March 2002 and operated until April 2012.

Across the 1990s, COSSA was active in a wide range of remote sensing projects. Between May 1991 and December 1992, Australia took part in the International Land Cover Project, an international environmental monitoring activity for the International Space Year. COSSA participated in this project to study the extent and causes of global land cover degradation using satellite remote sensing data to monitor and record change over time. COSSA consulted on the development of NASA's Spaceborne Imaging Radar (SIR) payload (which flew on two separate shuttle missions in 1994, STS 59 and STS 68) as well as the Japanese Advanced Earth Observation Satellite (ADEOS). The Office also identified a number of airborne remote sensors under development in various CSIRO divisions that had the potential to become satellite instruments. These included the Atmospheric Pressure Sensor (APS), the Ocean Colour Scanner (OCS), the Global Atmospheric Methane Sensor (GAMS) and an infra-red detector for atmospheric volcanic ash. However, COSSA was closed down before any of these sensors could be fully developed for space applications.

Under the reorganisation of responsibilities for Australian space activities instituted by the Howard Government in 1996, COSSA was restructured to undertake the FedSat technology demonstrator satellite project, with its core staff transferred to a new Cooperative Research Centre for Satellite Systems (CRCSS) (see Chapter 8). The CSIRO Earth Observation Centre (EOC), which had been established within COSSA in 1995, absorbed the remaining COSSA staff and continued with remote sensing research activities until it, too, was closed in 2004. Inheriting some of the responsibilities of COSSA and EOC, CSIRO Space Sciences and Technology was established in 2008, although it was not a direct replacement for the earlier groups. In late 2009, a new division, CSIRO Astronomy and Space Science (CASS), was formed, bringing together CSIRO's radio

Above (Top) Australia was a co-owner of the Advanced Along Track Scanning Radiometer instrument flown on Envisat, the successor to the ERS spacecraft. The satellite operated between 2002 and 2012, producing significant environmental data. (Image courtesy of ESA)

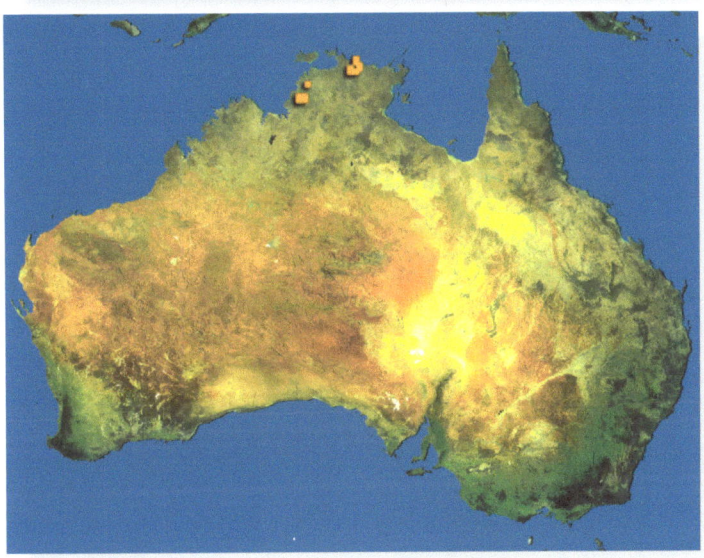

Above (Bottom) Fire spots detected in Australia on 20 June 2006, by the Advanced Along Track Scanning Radiometer instrument on ESA's Envisat Earth observation satellite. (Image courtesy of ESA)

astronomy capabilities (the Australia Telescope National Facility), NASA Operations (including the Canberra Deep Space Communication Complex) and CSIRO Space Sciences and Technology: it remains the present CSIRO section with responsibility for space science, although its status was changed in 2014 from that of a 'Flagship' division to a 'business unit', signifying a reduced priority in CSIRO's research activities. In 2015,

CSIRO and ESA entered into a new agreement for collaboration between Australian and European researchers in the evaluation of satellite data for use in Australia, while jointly developing new applications and space technologies for future satellites.

The Aggregation of Red Cells Experiment: Medical Research in Orbit

Without support from either the ASO or COSSA, another Australian payload was flown twice on the Space Shuttle in the 1980s. The Aggregation of Red Cells (ARC) experiment was the brainchild of Dr Leopold Dintenfass, a rheologist and medical researcher with the University of Sydney and the Rachel Forster Hospital. From the late 1970s, Dintenfass fought for funding for this project, which was designed to examine the effect of the microgravity environment of space on the aggregation of normal and diseased human red blood cells, to determine if any differences of reaction could provide keys to treatment or cure for the diseases being studied. With little support from his university for his project, Dintenfass raised money from a variety of sponsors, including real estate company Jones Lang Wootton. The company agreed to support the ARC experiment when Dintenfass contacted them on the basis of an advertising campaign that they were undertaking, which used the slogan 'We put more people into space than the Americans and Russians combined'. Dintenfass convinced them to put their money where their slogan was.

The ARC experimental apparatus was designed by Dintenfass and built by Hawker de Havilland in Sydney. The apparatus operated by passing blood samples through slit-capillary photoviscometers (devices for measuring the viscosity, or fluidity, of blood) developed especially for the experiment. The blood samples in space were photographed simultaneously with control samples on the ground. Because the blood samples used in the experiment had to be fresh, volunteer donors paid their own way to the US in order to be available when required prior to the experiment's launch.

Space Shuttle flight STS 51C carried the ARC experiment aloft for the first time on January 1985. The experimental apparatus was designed to be carried in a locker on the Shuttle's Mid-deck. Post-flight analysis of the experiment showed that aggregation of red cells does occur in microgravity conditions, with apparently different results in healthy and diseased bloods.

Given the potential importance of such a discovery, NASA offered Dintenfass an early re-flight of his instrument and the experiment flew again on STS 26 in 1988, the first Shuttle flight after the loss of the Challenger. NASA was so pleased with the results of this flight that the experiment was even scheduled to make a third trip into space. However, Dintenfass passed away suddenly in 1990 and the project immediately came to a halt. Despite efforts to find another researcher to take up the project, none expressed interest and the ARC experiment did not fly again: the research remains incomplete.

Below Dr Dintenfass prepares the Aggregation of Red Cells experiment for its first Shuttle flight at Kennedy Space Centre. (Photo courtesy of NASA)

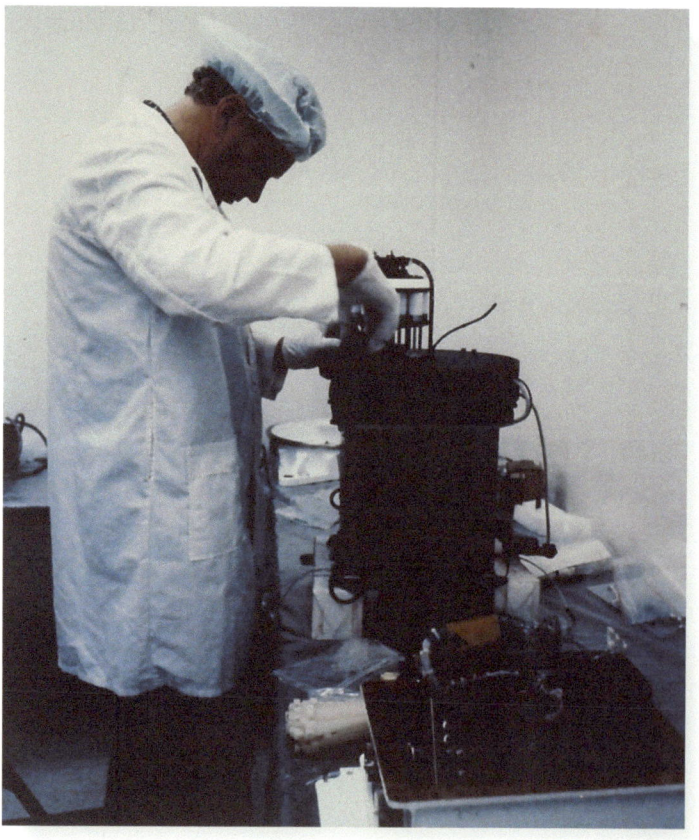

LINKING SPACE TRACKING AND RADIO ASTRONOMY: VLBI AND RADIO SCIENCE-BASED RESEARCH

When NASA's space tracking facilities were initially established in Australia, the agreement between Australia and the US included a provision that the antennae could be used for astronomical research when not required for tracking duties. Australian and international astronomers have utilised the Australian tracking stations since the 1960s, in particular the Deep Space Network stations at Island Lagoon and Tidbinbilla, for a wide range of radio astronomy research.

Radio science research, which uses radio frequency experiments between spacecraft and the Deep Space Network's antennae, have allowed planetary scientists and astronomers to uncover and better understand many characteristics of the Solar System and also to confirm general relativity. The powerful transmitting capabilities of large tracking antennae and their ability to detect extremely weak radio signals have been used for radio and radar astronomy projects in astrophysics, geophysics and planetary radar studies, in which radar signals are reflected off the surface of planets or other celestial bodies to determine their surface features.

Tidbinbilla has also participated in Earth dynamics studies, which use the GPS system to help refine precise position measurements on the Earth's surface, and 'sky surveys', which involve the identification and mapping of radio sources to create radio position reference frames.
An area in which the Tidbinbilla tracking station and Australian radio astronomy facilities have made important contributions is that of Very Long Baseline Interferometry (VLBI). The technique of VLBI involves linking radio telescopes many kilometres apart, so that they effectively act as a giant single radio telescope, with a diameter equal to the distance across the array of telescopes. This greater diameter means that higher resolution can be achieved in astronomical observations.

Linking ground-based facilities with radio telescopes in orbit creates even more powerful 'virtual' radio telescopes, and projects in this

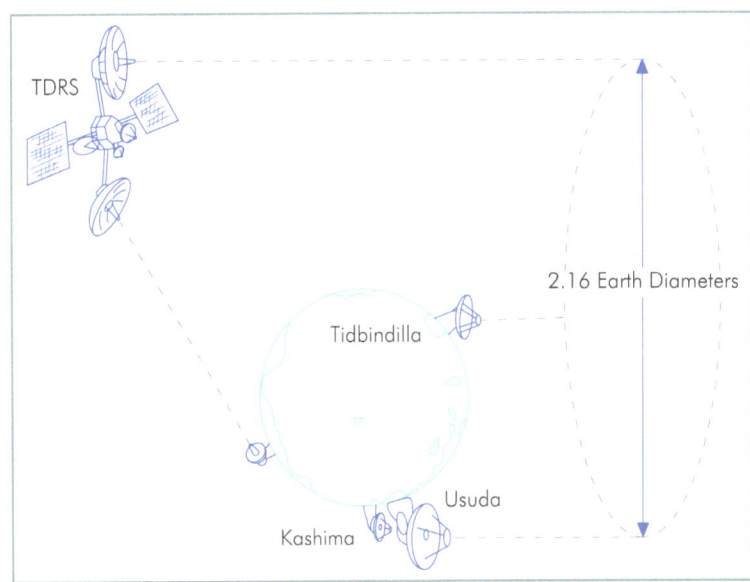

Above Diagram showing how Tidbinbilla tracking station linked with Japanese radio telescopes and a NASA TDRS satellite to create a Very Long Base Array virtual telescope with a baseline 2.16 times the diameter of the Earth.

area commenced in the 1980s, with Australian astronomers being involved in the early planning stages of several space VLBI satellite projects, including the Russian RadioAstron mission. Although greatly delayed, RadioAstron was eventually launched in 2011 carrying a 20cm radio receiver designed and supplied by CSIRO. In 1986 and 1988, Tidbinbilla and Japanese radio telescopes were used in conjunction with NASA's Tracking and Data Relay Satellites (TDRS) to create an interferometer with a baseline 2.16 times the diameter of the Earth to study remote radio sources in deep space. This was a precursor to Japan's VSOP (VLBI Space Observatory Program, also known as HALCA) satellite, which operated between 1997 and 2005. NASA's Deep Space Network created a subnetwork of 11-metre antennae to support VLBI research in the late 1990s.

The establishment of a permanent microwave link between Tidbinbilla and the Parkes Radio Telescope in 1986, to enable Parkes to support the Voyager encounters with Uranus and Neptune, created the world's fastest real-time VLBI interferometer. It also opened the way for VLBI experiments linking the Australia Telescope Compact Array at Narrabri, NSW (opened in 1988), the Parkes Radio Telescope, Tidbinbilla, the

How Does a Hypersonic Shock Tunnel Work?

A shock tunnel consists of two long tubes end to end, with one of a larger diameter than the other. The larger diameter compression tube contains a free piston and is separated by a metal diaphragm from the second, smaller diameter shock tunnel. At the end of the smaller tube, a nozzle and test chamber are attached. In tests of aerodynamic models, the piston is driven along the compression tube at speeds up to 300 metres per second. The head of the piston compresses the helium gas, with which the tube is filled, so much that the diaphragm separating the compression tube from the shock tunnel bursts. Helium then flows into the shock tube displacing the test gas (usually air) with which it was filled. A shock wave travels down the tube, creating a pocket of test gas, which expands through the nozzle at the end of the tube at hypersonic velocities and flows into the test chamber containing an airframe or engine model.

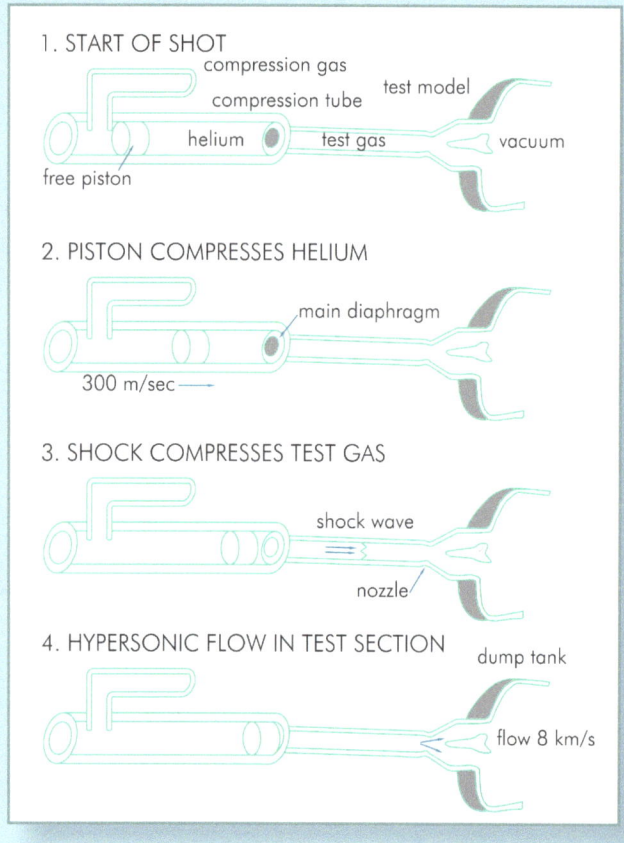

ACRES receiving station and the ESA satellite control centre at Gnangara in order to create a vast VLBI array. Stemming from those experiments, today the Australian Long Baseline Array (LBA) comprises the Australia Telescope Compact Array, Parkes, Tidbinbilla, and radio telescopes at Mopra (near Coonabarrabran, NSW), Hobart and Ceduna. The facilities can be flexibly linked in various configurations to suit particular research requirements and can also work in conjunction with overseas antennae to create even larger VLBI virtual telescopes.

Hypersonic Research: Higher, Further, Faster, Cheaper

One of the main reasons for the rocket-based hypersonic research projects conducted at Woomera in the 1950s and 60s (Chapter 5) was that the wind tunnels and computing technology of the day were not advanced enough to permit simulation research at hypersonic velocities. However, to reduce the need for expensive rocket trials, Australia led the world in developing hypersonic wind tunnels in the 1960s and has remained at the forefront of this technology ever since.

Hypersonic wind tunnels, capable of generating very high-speed airflows in the Mach 5 to 25 range, enable real gas simulations on the ground of the high altitude conditions encountered by aircraft travelling at hypersonic speeds and spacecraft re-entering the Earth's atmosphere. The only technology capable of producing the necessary high-speed airflows is the free-piston shock tunnel, which was invented in Australia in the early 1960s by a group at the Australian National University.

Australian aeronautical engineer Ray Stalker pioneered the development of the free-piston shock tunnel, with three progressively improved models, designated T1 through T3, constructed at the Australian National University in the 1960s and early 70s. They ranged in length from three metres to 12 metres. In 1977, Stalker moved to the University of Queensland, initially as Professor of Mechanical Engineering before becoming Australia's first Professor of Space Engineering in 1988. At Queensland, Stalker constructed what was, at the time, the world's most advanced shock tunnel, the 35-metre long T4. Since 1987, this facility has played a fundamental research role in the development of hypersonic SCRAMJET technology, another field in which Stalker was a pioneer.

The SCRAMJET, or supersonic combustion ramjet, is an air-breathing engine that draws oxygen from the atmosphere, and forces it at supersonic velocities into a combustion chamber, where it is burned with fuel to produce thrust. Unlike a conventional jet engine that uses turbines to force air into the engine, a SCRAMJET uses the speed of its travel to force air directly into the combustion chamber. To achieve this, it must be travelling at velocities between Mach 5 and Mach 25, which means that a SCRAMJET-powered vehicle requires some other form of propulsion to achieve hypersonic velocity. However, operating without heavy turbines and fans means that a SCRAMJET-powered vehicle is not only fast, but also comparatively light, requiring less fuel than either a conventional jet engine or a rocket.

Real progress in SCRAMJET technology began in Australia, with Stalker undertaking early SCRAMJET experiments in 1981 using the T3 shock tube at ANU. The 1980s saw a surge of interest in SCRAMJET technology, which was seen as having the potential to power both the middle stages of satellite launch vehicles (while the rocket was still within enough atmosphere from which the engine could draw its oxygen) and sub-orbital spaceplanes that could intercept military threats from space or transport passengers around the world in a few hours (reducing the travel time from Sydney to London, for example, to two hours). Their lower fuel requirements also meant that a SCRAMJET-powered vehicle would be cheaper to operate than conventional rocket stages or supersonic aircraft, making them attractive alternatives to conventional launch vehicles.

Space agencies and aerospace companies investigated aerospace plane designs and advanced re-entry systems, many of which were tested at Australia's hypersonic research facilities. The ANU T3 facility undertook testing for the British Aerospace single stage to orbit vehicle HOTOL and the European Space Agency's Hermes spaceplane. The University of Queensland's T4 shock tube, which was specifically designed to test full-scale SCRAMJET modules, was used for research on NASA's National Aerospace Plane (NASP) and the development of the Re-Entry Air Data System (READS) for the Space Shuttle. In 1991, Stalker proposed an Australian SCRAMJET-powered launcher, the Queensland Mach 20, which would utilise local SCRAMJET expertise and aviation skills to build a lightweight SCRAMJET vehicle as the forerunner to a future spaceplane.

While aerospace planes proved more difficult to develop than was originally envisaged, and all the early proposed vehicles were subsequently cancelled, SCRAMJET research continued at the University of Queensland, which, in 1993, achieved what is arguably the world's first 'flight' of a SCRAMJET, using a scale version of a prototype engine tested in the T4 tunnel. This

BELOW Shockwaves surrounding a test model of the Space Shuttle Orbiter in the T4 shock tunnel. This research contributed to the development of the Re-Entry Air Data System (READS) for the Space Shuttle. (Image courtesy of the Centre for Hypersonics, University of Queensland.)

Above Launch of the HIFiRE 5b SCRAMJET test at Woomera in 2016, carrying an experimental SCRAMJET engine from the University of Queensland Centre for Hypersonics. (Photo courtesy of Defence Science and Technology Group)

test model achieved a velocity over Mach 7 and demonstrated that it could develop the forward thrust that would produce acceleration in flight, thus proving the validity of the design. Building upon this success, the Centre for Hypersonics was established in 1997 with support from NASA. As described in Chapter 5, the Centre initiated the HyShot project to further the development of SCRAMJET technology, leading to five test flights being carried out at Woomera, with a range of defence and civil partners, between 2001 and 2007. The Centre also remains involved with HyShot's defence-focused follow on program, HIFiRE.

In 2010, the Centre for Hypersonics' SCRAMSPACE project became the first and largest project funded by the Australian Space Research Program, which awarded $5 million toward a total project cost of $14 million. Undertaken by an international consortium of partners in Australia, Germany, Italy, Japan and the USA, this project looked to conduct a flight-test of a 1.8-metre-long free-flying SCRAMJET at Mach 8, in addition to ground-tests aimed at achieving up to Mach 14. An additional goal was building capacity and capability for the Australian space and aerospace industry, through the training of a new generation of aerospace engineers.

Design, development and testing of the flight experiment payload were undertaken by the University of Queensland, with major support from DSTO, which co-ordinated the launch campaign. The University of New South Wales and the Italian Aerospace Research Centre provided additional flight experiments, while industry partners BAe Systems and Teakle Composites provided flight hardware, and pre-flight testing. Other university and space agency partners also contributed to the program, along with the Australian Youth Aerospace Association.

The launch took place at the Andøya Rocket Range in Norway in September 2013. Due to a failure of the first-stage rocket motor, the SCRAMJET payload did not reach the correct altitude and speed during its test flight, with the SCRAMJET crashing back into the water off the nearby coast. Although the flight test did not deliver hypersonic flight data, the project still produced important data from its ground test, modelling and analysis phases and paved the way for the University of Queensland's most recent program, SPARTAN, which aims to construct a reusable satellite launching system using hypersonic technology.

Hypersonic and SCRAMJET research using the T3 shock tube also continues at the ANU, with a long-term collaboration between ANU's Aerophysics and Laser-Diagnostics Research Laboratory and the University of New South Wales at ADFA. Using laser-based diagnostic techniques, several advances have been made in the understanding of atmospheric entry near-wake flows, including the first measurements of vorticity and velocity in these flows and new methods for visualising these flows that are more than 100 times more sensitive than the previous state of the art. In 2010, this team successfully demonstrated that the high-speed flow in a naturally non-burning SCRAMJET engine can be ignited using a pulsed laser source.

Another Way to Reach for the Stars: Plasma Propulsion

As chemical propulsion systems rely on finite quantities of fuel and oxidiser carried on board the rocket, their operational lifespan is limited to how much fuel they can carry. The search for alternative means of propulsion, which produce thrust from other sources than the chemical reaction of fuel and oxidiser, is another area in which Australia has led the world in recent decades.

Plasma thrusters, which eject high velocity ionised gas to provide thrust, have been a focus of research at the Australian National University's Plasma Research Laboratory under Prof Rod Boswell for two decades. In 2003, Boswell and Dr Christine Charles created spontaneous current-free plasma double layers in the laboratory, reproducing a natural phenomenon that occurs in aurorae when charged particles released by the Sun hit the magnetic field of the Earth. This creates a boundary of two plasma layers. Electrically charged particles pick up energy as they travel through the layers of different electrical properties and Boswell and Charles realised that these accelerating properties could be applied to the creation of highly efficient spacecraft thrusters.

Further research led Dr Charles to the invention of the Helicon Double Layer Thruster (HDLT), a small rocket engine that uses electricity to produce ionised gas, which is accelerated to create thrust. The HDLT design is revolutionary in that it is purely based on naturally occurring physical phenomena, without any moving parts, thus eliminating the problems of other plasma thruster designs, as well as engine corrosion. Using solar energy to create a magnetic field through which the tenuous gases of space are passed to make a thrust-producing beam of plasma, an HDLT thruster could have an almost infinite lifetime.

The potential application of the HDLT thruster to greatly decreasing the costs of maintaining satellites and spacecraft in their orbits, or accelerating spacecraft to the planets, drew early support to the project from the European Space Agency, Auspace and the CRCSS. Former NASA astronaut Dr Franklin Chiang-Diaz, who has developed a different plasma engine design known as VASIMR (Variable Specific Impulse Magnetoplasma Rocket) has also worked with the ANU on refining plasma engine design.

In 2006, simulation research by ESA verified the feasibility of the thruster design and in 2010, a prototype 15cm diameter thruster, operated in low-magnetic field mode, underwent initial thrust testing. The potential of the HDLT thruster technology to be developed into a unique Australian space export resulted in it receiving a grant under the Australian Space Research Program and in 2016, the final thruster prototype underwent space simulation tests at the Advanced Instrumentation and Technology Centre (AITC) at Mt Stromlo. The development of the HDLT thruster continues at the Space Plasma, Power and Propulsion group within the ANU's Plasma Research Laboratory with strong promise for the future.

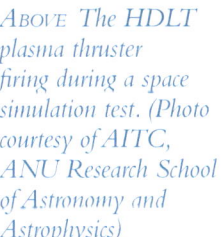

Above The HDLT plasma thruster firing during a space simulation test. (Photo courtesy of AITC, ANU Research School of Astronomy and Astrophysics)

Above HDLT plasma thruster prototype being mounted in a vacuum chamber at the Advanced Instrumentation and Technology Centre for space simulation tests. (Photo courtesy of AITC, ANU Research School of Astronomy and Astrophysics)

Astrobiology and SETI

Working across various universities and agencies, Australian geologists and other Earth scientists have made contributions to planetary studies, particularly in the multi-disciplinary field of astrobiology, where Australia is also considered at the forefront of current research to understand the origins of life on Earth and the possibility of life elsewhere in the universe.

Astrobiology is the study of the origin, evolution, distribution, and future of life in the universe: it seeks to identify whether life exists beyond Earth, and, if so, how it can be detected, through understanding how life first developed on our own planet. Australian geologist and palaeontologist Prof Malcolm Walter has been a pioneer in astrobiology since entering the field in 1987. In 2001, he founded the Australian Centre for Astrobiology (ACA) at Macquarie University, the first centre for astrobiological research in Australia and one of only two Associate Members of the NASA Astrobiology Institute outside the United States. In 2008, the Centre moved to the University of New South Wales.

The ACA is one of the few astrobiology research centres that is truly interdisciplinary and multi-disciplinary in a way that reflects the goals and aspirations of astrobiology as a scientific discipline. The Centre's goals include contributing to the understanding of what makes a habitable planet, studying the co-evolution of life and habitats on early Earth, and helping to guide the exploration for life elsewhere in the universe, particularly on the planet Mars, in which ACA researchers have a special interest.

Focusing on studies of stromatolites, an extremely ancient lifeform that can be found as both modern living specimens and in fossil form in Western Australia, the ACA conducts much of its research in the Pilbara region, which contains some of the world's oldest rocks and fossils. This research investigates the evolution of early life and its relation to changing habitats over time, the community structure of stromatolites and the diversity of cyanobacteria, which may have formed some of the earliest life on Mars as well as on Earth.

The ACA is also involved with the search for exoplanets and habitable worlds and its members and former students have played a variety of operational and investigative roles in NASA's Mars rover missions. In 2017, ACA researchers announced the discovery of the oldest evidence of life on land, pushing back the date of land-based life by around 580 million years. The finding suggests that life may have arisen in land-based hot spring environments rather than in hydrothermal vents deep in the sea, a discovery that has implications for the search for past life on Mars: exploration of ancient hot spring sites on Mars could yield fossilised traces of the earliest Martian life.

Above Living fossils. The modern stromatolites of Shark Bay, Western Australia, are a survival of one of the oldest life forms on Earth. By studying modern and fossil stromatolites, astrobiologists hope to gain insight into the possibility of early life on Mars. (Author's collection)

In addition to its scientific research, the ACA is also at the forefront of studies in science communication and education and operates a world-leading media, education and outreach program related to this research. This includes a Mars Yard at the Powerhouse Museum in Sydney, the development of which was funded by the first education grant awarded under the Australian Space Research Program. The ACA has also developed, in association with NASA, a 'virtual field trip', allowing students to join scientists in ACA field study areas in Western Australia, where research is being conducted on both modern stromatolites and the ancient stromatolite fossils that are the oldest convincing evidence of life on Earth.

While astrobiology is concerned with the possibility of life of any kind elsewhere in the universe, the Search for Extraterrestrial Intelligence (SETI) is particularly focused on finding evidence of intelligent life beyond the Earth, through the detection of radio or other signals that could only have been produced by a technological civilisation.

Australian radio astronomers first undertook a SETI search program in 1979, using the Parkes Radio Telescope. Unlike the usual SETI searches, focused on finding narrow-band beacons around the so-called 'hydrogen line', which exists in the quietest part of the radio spectrum, this search was directed toward seeking wide-band pulsed signals, an idea that was considerably ahead of its time. They detected only one signal that could not be explained as either a natural source or radio frequency interference (a common cause of 'false' SETI detections). However, a repeat of the signal was never detected and it remains unexplained.

In 1983, DSS 43, the largest antenna at the Tidbinbilla tracking station, was used for SETI observations with funding support from the Planetary Society, the largest US space advocacy association. Another SETI search was conducted at Parkes in 1989. This radio telescope has become the backbone of Australian SETI research, with almost all studies to date conducted there. For the 1989-91 search, researchers targeted 100 nearby stars similar to our Sun, many of which are only visible from the Southern Hemisphere. One signal was detected that was not natural and lasted long enough to determine its sky location. However, repeated observations of the location were unsuccessful in detecting another signal.

Following the cancellation of NASA's SETI research in 1992, the SETI Institute of Mountain View, California, took over its SETI program and funded new research under the name Project Phoenix. The Southern Hemisphere component of this search took place at Parkes in February and June of 1995. Phoenix concentrated on nearby star systems similar to our own, investigating approximately 800 stars within a range of 200 light years. In March 2004, Project Phoenix announced that it had failed to find any evidence of extraterrestrial signals. An additional short SETI search, using the Project Phoenix equipment at Parkes, also took place in June 1995, aimed at the Small Magellanic Cloud in the hope that its large stellar population would provide a better chance of detection than targeting individual stars within the Milky Way.

In the mid-1990s, the SETI Australia Centre was established at the University of Western Sydney's Cambelltown campus. In 1998, the Centre inaugurated its Southern SERENDIP project at Parkes, as the Southern Hemisphere counterpart to the SETI Institute's SERENDIP project. Jointly funded by the SETI Institute and the University of Western Sydney, Southern SERENDIP observations continued until 2009, although the data still remains largely unprocessed.

The same year that Southern SERENDIP commenced, the first optical SETI (OSETI) experiment in Australia was carried out at the Perth Observatory in Western Australia. Optical SETI is based on the suggestion that advanced extraterrestrial cultures may use laser beams for communication across space, rather than radio, and employs optical telescopes to search for pulses of laser light in or near the visible portion of the spectrum. An OSETI project was commenced in 2000 at the University of Western Sydney. In December 2008 a sharp laser look-alike signal was detected, apparently emanating from 47 Tucanae globular cluster, but further searches failed to detect the signal again.

In 2007, the first VLBI SETI search was undertaken using the Australian Long Baseline Array, and in 2014, the Murchison Widefield Array, a technology and science pathfinder for the Square Kilometre Array radio telescope to be constructed across Australia and South Africa, was used to search for narrowband transmissions from 38 known planetary systems. Most recently, in 2016, the Breakthrough Listen project, co-ordinated by the University of California, Berkeley, commenced a new SETI search at Parkes.

Although no SETI project has yet detected a verifiable signal of intelligent extraterrestrial origin, the hope remains that perhaps such a world-changing discovery might one day be made using an Australian radio telescope.

The breadth and depth of Australian space science and engineering is indicated by the projects highlighted in this chapter, but these achievements should be recognised as only being a part of the overall contribution made by Australian scientists and engineers working in government, industry and the universities both in Australia and overseas. Many have contributed to attempts to develop an Australian space industry that would benefit the country both economically and by reducing our reliance on space-based goods and services provided by other countries: their story is continued in the following chapter.

'Australian' Astronauts

Since Yuri Gagarin became the first person in space, in 1961, astronauts and cosmonauts have been considered national status symbols. Although no Australian citizen has yet flown in space wearing the flag of the country on their flightsuit, there have been three Australian-born space travellers who had become US citizens prior to their selection by NASA.

Below Australian-born NASA scientist-astronaut Dr Philip K Chapman. (Photo courtesy of NASA)

Dr Phillip K Chapman

The first Australian-born person to be selected as an astronaut was Dr Phillip K Chapman, originally from Melbourne. NASA chose Chapman as part of the second group of scientist-astronauts recruited for the Apollo program in August 1967: he was one of the first two naturalised US citizens to be selected as an astronaut, having become an American citizen only a few months earlier. A physicist and a specialist in instrumentation, Chapman had studied at the University of Sydney and the Massachusetts Institute of Technology (MIT), from which he obtained a degree in aeronautics and astronautics.

Prior to being selected as an astronaut, Chapman had studied aurorae in Antarctica, as part of the Australian expedition there during the International Geophysical Year. He also worked on aviation electronics in Canada before joining MIT as a staff physicist in 1961. Employed in MIT's Experimental Astronomy Laboratory at the time of his selection as an astronaut, Chapman had previously, in 1964, put forward a proposal to the Australian Government for the establishment of an Australian-managed international equatorial launch facility on Manus Island, off Papua New Guinea, as noted in Chapter 3.

Chapman's selection made him part of a special group of 11 scientist-astronauts whom NASA intended at that time to participate in the Apollo Applications Program, which was to have been the follow-on from the initial Apollo lunar landing program. This program was to have included an additional ten Moon landings and three orbiting research laboratories. However, cutbacks in the NASA budget over the next few years reduced the Apollo follow-on program to the single Skylab space station, and Chapman resigned from NASA in 1972 when it became clear that, despite serving on the support

Born in Sydney in 1944, Scully-Power graduated from the University of Sydney with a degree in education and applied mathematics and established the Royal Australian Navy's first oceanographic group in 1967. Scully-Power headed this unit until 1972, when he went to the US as an exchange scientist at the US Navy's Underwater Systems Centre in Connecticut. While working there Scully-Power became involved with briefing and training the Skylab astronauts for oceanographic observations. In 1975, he was again involved in the briefing of astronauts for the Apollo-Soyuz Test Project.

After acting as Principal Investigator on an experiment on the 1976 Heat Capacity Mapping Mission satellite, Scully-Power became a US citizen in 1977 and took up a permanent post as a senior scientist with the Underwater Systems Centre. He was also a flight crew instructor in the NASA Astronaut Office at Houston between 1980 and 1986.

In June 1984, he was selected to fly on the STS 41G Shuttle mission when the reassignment of that mission from the Shuttle Columbia to the Shuttle Challenger (which had one additional crew seat) allowed the addition of another Payload Specialist to the crew, to take advantage of that mission's unusual orbital inclination of $57°$. This provided a much wider view of the Earth than the more usual Shuttle orbital inclination of $28°$.

Above Oceanographer Dr Paul Scully-Power in training for his flight on the Space Shuttle as a Payload Specialist in 1984. (Photo courtesy of NASA)

During his flight, Scully-Power conducted visual oceanographic observations of three-quarters of the world's oceans. Because of the height of the Shuttle's orbit (200-400 kilometres), Scully-Power was able to observe ocean current features that could not be detected from the much higher orbiting remote sensing satellites. He made the important discovery that spiral eddies (circular ocean currents about 8-48km across) were not the rare ocean features they had previously been thought to be. Scully-Power's observations demonstrated that spiral eddies are common throughout the world's oceans, occurring in complex fields of interconnected systems extending for hundreds of kilometres.

crew for Apollo 14 in 1971, he would have no opportunity to fly in space until the Space Shuttle came into service, which would not occur until 1981.

Dr Paul Scully-Power

The first Australian-born person to actually make a spaceflight was oceanographer Dr Paul Scully-Power, who served as a Payload Specialist aboard the STS 41G Space Shuttle mission in 1984. During the Space Shuttle program, a Payload Specialist was not a permanent member of the astronaut corps and received a spaceflight assignment on the basis of being an expert necessary to conduct specific experiments or research aboard the Shuttle.

'Australian' astronauts

Since atmospheric and oceanic systems are directly interrelated, improved understanding of ocean eddies enables a better understanding of large-scale ocean movements and their effects upon the world's weather systems, ultimately improving the accuracy of long-term global weather forecasting. Consequently, the discoveries made by Scully-Power during his space mission were of major importance to oceanography, meteorology and other Earth sciences.

A strong proponent of Australian participation in space activities, in 2001 Scully-Power was Chair of the International Space Advisory Group, appointed by the Howard Government to identify opportunities for Australian involvement in the International Space Station (ISS) and other international space programs.

BELOW *Spiral eddies in the Mediterranean Sea, photographed by Paul Scully-Power during his STS 41G Shuttle mission. (Photo courtesy of NASA)*

Dr Andrew Thomas

In 1992, Australian-born aerospace engineer Dr Andrew SW Thomas was chosen to become an astronaut. A specialist in microgravity research, he was selected as a Mission Specialist: during the Space Shuttle program, Mission Specialists were permanent members of the astronaut corps responsible for the various experiments and research programs carried out on Shuttle missions.

Born in Adelaide in 1951, Dr Thomas obtained both his bachelor's degree and his PhD in mechanical engineering from the University of Adelaide. In 1977, he joined the Lockheed Aeronautical Systems Company, based in Georgia in the US, where he worked until 1989 on a variety of aerodynamics research projects. In 1986, Thomas became a US citizen and in 1989 he joined the technical staff of NASA's Jet Propulsion Laboratory (JPL), where he gained considerable experience in microgravity (weightlessness) studies. This was an important factor in his astronaut selection, as NASA planned a major focus on microgravity research in the Space Shuttle science programs of the 1990s and also on the then-in-planning International Space Station.

In his astronaut career, Thomas made four spaceflights, totalling just over 177 days in space. His first flight was STS 77 in May 1996, aboard the Space Shuttle Endeavour. This mission included the deployment of two satellites, testing a large inflatable space structure on orbit and a number of scientific experiments conducted in a Spacehab laboratory module.

Changes in US citizenship laws enabled Thomas to reapply for Australian citizenship in late 1997, and he made his second spaceflight as a dual Australian-US citizen, thereby becoming the first Australian citizen in space, while still serving as an American astronaut. In January 1998, Thomas launched on STS 89, which delivered him to the Russian space station Mir, where he would serve for the next four months as Flight Engineer 2, the last US astronaut participant in the Shuttle-Mir program, which was designed to extend NASA's experience in long-duration spaceflight in advance of the International Space Station program. He returned to Earth aboard Space Shuttle Discovery in June 1998, during the STS 91 mission, having spent 141 days in space.

ABOVE *Dr Andrew SW Thomas, NASA Mission Specialist astronaut, photographed in an EVA spacesuit. This official portrait was taken prior to the STS-102 mission during which Thomas made his first spacewalk. (Photo courtesy of NASA)*

STS 102, in March 2001, was Thomas' third flight, during which he performed his first EVA (Extravehicular Activity), or spacewalk, spending six-and-a-half hours installing components on the exterior of the ISS. Between 2001 and 2003, Thomas served as Deputy Chief of the Astronaut Office.

Thomas' final space mission was STS114, in July 2005. This was the Return to Flight mission following the Columbia accident, during which the crew continued the assembly of the International Space Station. Thomas tested and evaluated new procedures for flight safety, and inspection and repair techniques for the Shuttle's thermal protection system. Shortly before this flight, Thomas married fellow astronaut Shannon Walker. He retired from NASA in 2014.

A passionate supporter of Australian participation in space activities, Thomas has frequently spoken publicly to urge Australian Government re-engagement with space and was also a member of the International Space Advisory Group. He has strongly advocated for Australia to establish a national space agency and participate in the International Space Station program.

Although Andrew Thomas has flown into space as an Australian citizen, he did not fly as an Australian national astronaut. Although Australia was one of the first countries to express interest in NASA's 1985 offer to allied nations to train and fly a national Payload Specialist on the Space Shuttle, the offer was never taken up. There were proposals to fly a national Payload Specialist in conjunction with both the first Aussat launch and the planned Endeavour Bicentennial spaceflight in 1988, but neither came to fruition, and other Payload Specialist proposals foundered, as successive Governments were unwilling to fund the project.

BELOW *The first Australian citizen in space, Thomas exercises in the Krystall module on the Russian Mir space station during his long-duration spaceflight in 1998. (Photo courtesy of NASA)*

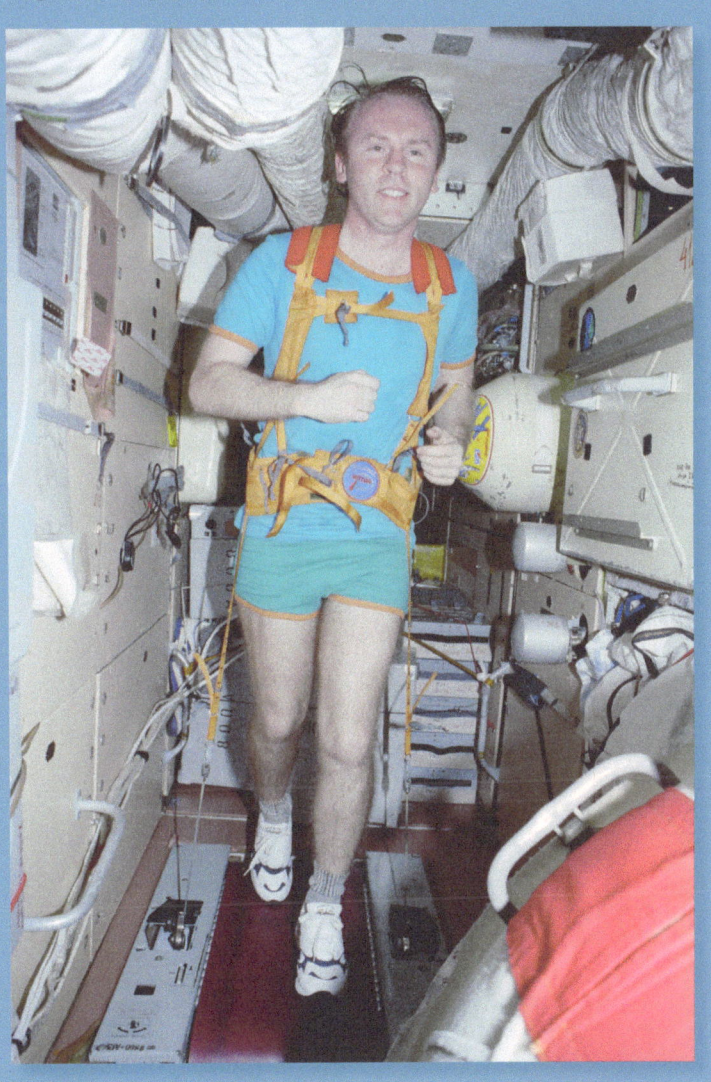

ABOVE *The logo for the EC0 cubesat project, one of the Australian satellites participating in the international QB50 scientific research program as a demonstration of Australian technology and engineering skills. (Courtesy of ACSER, UNSW)*

CHAPTER 8

Groundwork: Australian Space Industry

'Space industry' is a broad term that covers not only the manufacture of spacecraft, instrumentation and launch services, but also a wide range of ground sector activities that are integral to the operation and management of the space sector, together with applications that utilise space-enabled services. A 'space industry' can be any industry that utilises or contributes to the manufacture, operation or processing of space products, services and received data.

While Australia, even at its most 'space active', has never been more than a very minor player in the global space industry, such a categorisation belies the decades of effort that Australian engineers, scientists, commercial companies and research institutions have committed to attempting to develop a strong local space industry that would enable Australia's unique innovations and technical skills to be developed into profitable space products. While the development of an Australian space industry has waxed and waned with the cycle of Australian Government engagement and retreat from space activities, the spirit within the Australian space community that the country does have the potential to become a significant player in the global space industry has never wavered.

ABOVE *Australian designed and built antennae at OTC's Perth International Telecommunications Centre at Gnagara in 1990. (Photo courtesy of Telstra Corporation Ltd)*

BACKGROUND IMAGE OPPOSITE PAGE *The Advanced Instrumentation and Technology Centre at Mount Stromlo, Canberra, provides state-of-the-art facilities to support the manufacture of astronomical instruments and testing of space hardware. (Photo courtesy of AITC, ANU Research School of Astronomy and Astrophysics)*

Like the story of Australian space science and engineering outlined in Chapter 7, the full story of Australian space industry cannot be encompassed in a single chapter. Rather, this chapter will outline the broad trends and major events across the decades, highlighting particular projects that have had significant impact on the development of space industry in Australia.

ROCKETS AND WRESAT: EARLY AUSTRALIAN SPACE INDUSTRY

Australia's earliest space-capable industries developed in conjunction with the establishment and operation of the Woomera Rocket Range. British aviation and defence companies with missiles and weapons to be tested on the Range established local subsidiary companies to support their work. Most of these were based in South Australia, in proximity to WRE Headquarters at Salisbury. This early concentration of defence-related technical skills and expertise in South Australia lies behind that state's position today as the centre of Australia's defence industry. Even after the end of the Joint Program, many of these companies stayed on in Australia: some, like British Aerospace Australia and Hawker de Havilland, became important players in subsequent attempts to develop an Australian space industry.

Australian engineering and manufacturing companies also sought involvement with Woomera's establishment and operation, eager to develop a skilled workforce and gain technology that would bring their technical capabilities up to the cutting edge. The first company to be established at WRE Salisbury was, in fact, the Fairey Aviation Company of Australasia Pty Ltd, formed in 1948 as a joint venture between the British Fairey Aviation Company and Sydney-based Clyde Engineering. Local defence contractors acquired the capability not just to 'build to print' from foreign designs, but also to design and manufacture rocket motors, instruments, telemetry systems and ground station equipment. The Australian Defence Scientific Service developed significant in-house expertise in research, design and, to some extent, manufacturing of particular technologies, including optics, propulsion, tracking and telemetry systems and remote control systems for missiles and aircraft. By the late 1960s, Australia had significant space industry capabilities, honed through the Australian

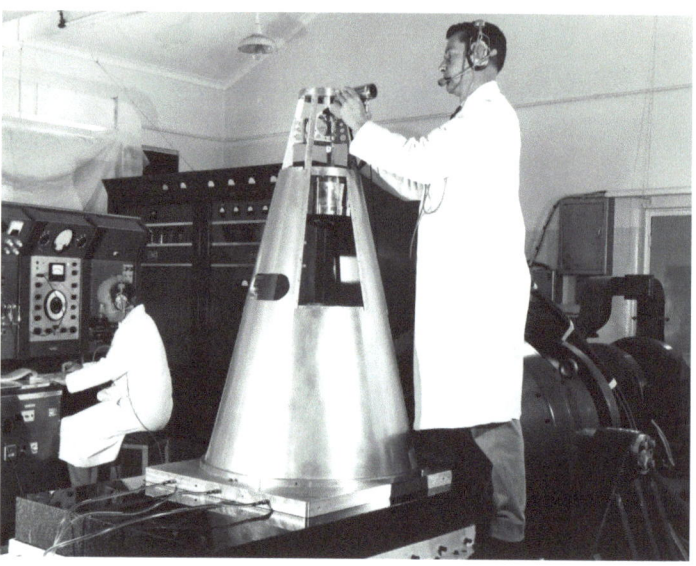

ABOVE WRESAT represented the pinnacle of Australian space science and industry in the 1960s. Despite its potential as the precursor of an Australian space program, there was no follow-on to WRESAT due to Government unwillingness to fund further development. (Photo courtesy of Defence Science and Technology Group)

sounding rocket program and the WRESAT and Australis-OSCAR 5 satellite projects, that could easily have been drawn together into a strong national defence and space industry. The WRE attempted to do this by presenting to Government in 1968, following the success of WRESAT, a plan for a combined defence and civil Australian space program. Although the program proposed was modest, it was declined on the basis of cost. The lack of the necessary seed funding required to initiate a significant Australian space industry building on local capabilities meant that, after the transfer of the ELDO project away from Australia and the demise of the Australian sounding rocket program, national space capability declined as the Joint Project wound down across the 1970s.

Although the Government drew back from supporting local space science and technology projects during the 1970s, thus causing established space capabilities to lapse, the increasing role that space applications, especially meteorology, remote sensing and communications, were assuming in Australia's national development spurred the development of capabilities applicable to the ground sector of space activities. These capabilities and Australia's general technology skills would enable the redevelopment of a small local space industry in the 1980s and 1990s.

THE CATCH-22 OF SPACE QUALIFICATION

By the beginning of the 1980s, the risk tolerance of the experimental years of the 1960s had faded and spacecraft owners and operators, especially in the commercialised satellite communications industry, wanted to avoid equipment failures that might cause the loss of a costly spacecraft. Consequently, 'space qualification' was necessary to prove the spaceworthiness of components for major spacecraft if a manufacturer had not already proven its reliability in previous successful spaceflights. It was a Catch-22 situation for many countries with emerging space programs: often an instrument or piece of equipment could not be flown until it was space qualified, yet it could not be space qualified without a spaceflight!

Because the country had been absent from space activities during the 1970s, it had effectively lost its 'space qualification' and Australian industry needed to prove that it was equal to the demanding standards of space engineering through an initial qualification flight. Involvement with the Starlab space telescope (and its subsequent Endeavour GASCAN payload) was seen as a means by which Australia could space qualify an example of local space technology and break the Catch-22 cycle.

THE 1980s: AUSTRALIAN SPACE INDUSTRY REVIVES

As outlined in Chapter 7, the Starlab space telescope project, coupled with the decision to develop the Aussat satellite system, was a catalyst for the revival of the Australian space industry. The 1980 Australian Space Industry Symposium drew more than 100 participants from fields as widely varied as battery manufacture and communications technology and revealed that many companies were fired by the challenge of breaking into the international space market. The fifteen companies that supported the Starlab project when funds were not forthcoming from Government for the research and development phase of the project collectively provided more than one million dollars of cash and in-kind support.

Aussat and the Offsets Program

The major expenditure for the establishment of the Aussat domestic satellite communications system was, as noted in Chapter 6, the purchase of three satellites and their delivery to orbit. The Fraser Government decided to encourage the development of a local space industry in order to recoup some of the Aussat expenditure, and pave the way for future national communications satellites to eventually be either partially or wholly constructed in Australia, through the Defence and Civil Offsets Program.

Originally set up to enable Australian industry to gain some benefit when capital expenditures were made overseas for high-cost items (such as defence systems and civil airliners), the offsets program required overseas suppliers to place work orders with Australian industry for components, sub-systems and services that could be provided locally. This system was a way of 'offsetting' the loss to local companies of a major contract being placed overseas: it was envisaged as a means of facilitating technology transfer, as well as providing business for local industry. Encouraging the growth of a local space industry through the Offsets Program was seen as a means of developing a local industrial and technical capability that would reduce the need for massive future expenditures on satellites, while also creating a future export industry. Although civil offsets for aerospace contracts would be phased out in 1991, they were an important contributor to the growth of Australian space industry in the 1980s.

Australia's lack of space-qualified industry meant that for the first generation of Aussat satellites, only cable harnesses and wiring systems for the satellites could be locally contracted. Australian experience and expertise in building satellite ground stations, however, did bring contracts for the Aussat ground segment. In addition, six Australian engineers (four from Aussat and two from the Government Aircraft Factories) were placed at the Hughes spacecraft construction facility in Los Angeles to participate in the Aussat satellite construction, to ensure that adequate technology transfer to Australia occurred. Despite early hopes that a significant industry would develop around Aussat, with growing Australian participation in the construction of the second and subsequent generations of satellites and the ground sector requirements of the system, the envisaged industry growth was not realised. CSIRO and a small number of local companies, such as Hawker de Havilland, British Aerospace Australia, Philips Australia, NEC Australia and Mitec, provided components for the Aussat/Optus B series and many invested heavily in new facilities to undertake this work, in expectation of the follow-on contracts that were anticipated as part of the Aussat offsets agreement. However, the privatisation of Aussat to Optus and the loss of the civil offsets provisions changed the situation markedly and further work did not materialise.

Starlab and Auspace

The demonstrated support from industry for STARLAB and the ACRES upgrade contributed to the Government's change in attitude toward establishing an Australian space industry. In August 1982, $3 million was allocated to ANU for Starlab studies, with a total of $8.3 million provided up to the cancellation of the project in 1984. Because the detector package being designed for Starlab

BELOW Australian engineers at the Hughes Aircraft Company satellite manufacturing facility in Los Angeles, where they participated in the construction of the first generation Aussat satellites to prepare for future satellite construction projects in Australia. (Photo courtesy of Optus Satellite)

was one of the most sophisticated instruments designed for a space telescope up to that time, it was regularly inspected by teams from NASA and Canada, passing each review with flying colours.

The Starlab project gave rise to Auspace, a local space company that would play a significant role in Australian space industry for twenty years. In 1981, French aerospace and defence company Matra expressed interest in the Starlab project and provided early technical assistance. This led, in 1983, to the formation of a joint venture called Auspace, in which Matra and the Mount Stromlo Observatory Starlab team (which included Hawker de Havilland) were 50/50 partners. Auspace became an independent space company in 1986, with the entire Starlab team leaving ANU to form its core and Matra as a major shareholder.

Auspace was funded to continue the studies for the instrument package for FUSE/Lyman, after the cancellation of Starlab, before Australian participation in that project, too, was cancelled by the ASO in 1988, for lack of funds. Instead, Auspace took a leading role in the Endeavour space telescope, investing heavily for the project in advanced manufacturing facilities for the development and construction of high-technology equipment, including a 'space-qualified' clean room, which was used for the construction and storage of Endeavour.

By the early 90s, Auspace had become a leader in the Australian industry, contracted to undertake the partial design and fabrication of the Focal Plane Assembly of the ATSR 2 remote sensing instrument for ERS 2. Auspace would also design and construct the Scan Mirror and associated control units for the Focal Plane Array of the Advanced Along Track Scanning Radiometer for the ERS successor satellite, Envisat. Among its other projects in the early 90s, Auspace developed a world-leading mobile GPS receiver that was later commercialised and marketed by another Australian company, Sigtec Navigation (now SigNav). These GPS receivers have been extensively used in space projects by the German Aerospace Centre (DLR).

Chapter 8 – Groundwork: Australian Space Industry

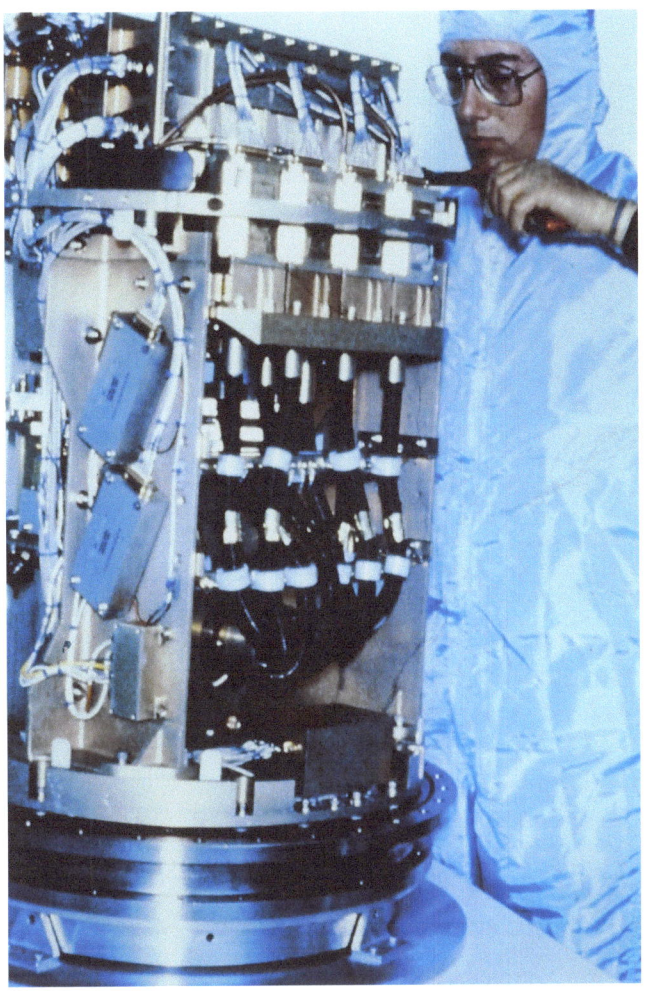

ABOVE *The Endeavour space telescope under construction in the Auspace clean room. (Photo courtesy of Auspace/Nova Group)*

OTC and the Telecommunications Ground Sector

Independent of the space industry growth around Aussat and Starlab, the 1980s also saw the rise of an active telecommunications ground sector industry, driven by OTC and the CSIRO, to meet Australia's particular telecommunications requirements. As satellite technologies improved, allowing the use of smaller satellite dishes, OTC, working in conjunction with CSIRO, designed and manufactured a wholly Australian 18-metre antenna, which was employed throughout its ground station network. OTC also developed a series of antennae that utilised Australian designs integrated with the most suitable components from a range of national and international manufacturers: ranked first technically and operationally by INTELSAT, the installation of these antennae at its Gnagara facility was instrumental in securing new TTC&M contracts for OTC.

The combined expertise of OTC and CSIRO in antenna design and ground station systems engineering, project management and operations enabled OTC to win international contracts to supply ground stations. By the early 1990s, OTC had provided ground stations in Laos, Cambodia, Vietnam, Malta, Kazakhstan, Antarctica and various Pacific islands, as well as Very Small Aperture Terminal (VSAT) data networks in Thailand and the Philippines. OTC also designed and managed the Pacific Area Cooperative Telecommunications (PACT) network, designed to assist in maintaining communications between various island nations with low telephone usage. This growing international presence would form an important component of OTC/Telstra's business strategy following the loss of its monopoly telecommunications provider status with the privatisation of Aussat, INTELSAT and INMARSAT, as outlined in Chapter 6.

Remote Sensing Image Analysis

Another area of space industry development in the 1980s was that of remote sensing image analysis. To facilitate the growing use of remote sensing imagery by the CSIRO, by the late 1970s its Image Systems Group had begun working on the development of software packages for image analysis and display with mini-computers as well as mainframes. One early success was DISIMP (Device Independent System for IMage Processing), a package for general purpose image handling, which was successfully commercialised and marketed in the 1980s. Other early software systems that were successfully marketed in Australia and internationally included Software for Landsat Image Processing (SLIP), designed to process Landsat and other satellite imagery and IMAGED (Image-based Analysis of Geographic Data), a set of programs designed to integrate, analyse and display different kinds of geographically referenced data. From these initial systems, special-purpose, commercially successful applications were developed, such as CSIDA (CSIRO System for Interactive Data Analysis), which was designed for use with high-volume daily imagery, such as that from NOAA and Japanese GMS weather satellites.

With the introduction of micro-computer-based work stations by the late 1980s, software specifically designed for image processing

Above MicroBRIAN was a highly successful software package for image processing on micro-computer workstations, developed by the CSIRO in the 1980s. Originally developed for marine research, MicroBRIAN was marketed for use with a wide range of remote sensing applications. (Photo courtesy of CSIRO)

and the manipulation of remote sensing data on these smaller systems was developed. The most commercially successful of these was MicroBRIAN, adapted from the CSIRO's Barrier Reef Image Analysis (BRIAN) software package, originally designed for marine research on the Great Barrier Reef. This versatile software package could be used for a variety of Landsat applications, in addition to marine research.

The upgrade of the ACRES facility at the beginning of the 1990s, to receive the complex radar data from ERS 1 and other planned remote sensing satellites, led to the development of an extremely compact (for the day), and relatively cheap, data processing hardware system, AETHERS (Advanced Equipment to Handle ERS). The heart of the AETHERS system was an Australian-designed supercomputer called the Fast Delivery Processor (FDP), developed by British Aerospace Australia. AETHERS' data processing capability could deliver an image in two minutes, compared to the more usual four hours with other systems. Another Australian remote sensing product successfully marketed internationally, AETHERS was adaptable to other high-speed data processing applications such as high-resolution medical imagery and military surveillance.

SPACE INDUSTRY AND THE NATIONAL SPACE PROGRAM

The interest that Starlab generated in the development of a national space industry prompted the Commonwealth Department of Science and Technology to convene the first National Space Symposium in March 1984, to better understand the opportunities for Australia that might stem from Government support for a space industry. As already described in Chapter 7, the Symposium gave impetus to the formation of the CSIRO Office of Space Science and Applications (COSSA), which in turn gave an immediate boost to Australian space industry through its participation in the development of the Along Track Scanning Radiometer instrument.

A further outcome of the symposium was a growing call from the space community for Australia to develop a comprehensive national space policy and a national space agency. The Hawke Government responded by commissioning the Australian Academy of Technological Sciences to investigate the potential of space science and technology for Australia. The resulting report, *A Space Policy for Australia* (also known as the Madigan Report) was released in 1985. This report, discussed in more detail in Chapter 9, recommended the establishment of a national civil space policy emphasising support for a space industry focused on remote sensing and the ground infrastructure sector. It also recommended the commitment of a high level of funding toward suitable space industry development projects over a five-year program.

The Hawke Government's response to the Madigan Report was to establish, in 1986, the Australian Space Board (ASB) to oversee a National Space Program (NSP). The ASB was not, however, a statutory authority as the Madigan Report had recommended, but rather an advisory body, with the role of facilitating the development of Australian space science and technology through identifying industry opportunities and establishing collaborations with international partners.

The location of the ASB within the then-Department of Industry, Technology and Commerce (DITAC) made clear the Government's intention to focus the National Space Program toward industry development, rather than establishing a broader program that would encompass all aspects of space science and engineering, the growing space applications sector (covering communications and meteorology as well as remote sensing) and the Defence space sector. Australian companies' lack of experience in space project management hampered their ability to compete for space contracts overseas. Consequently, the National Space Program was specifically aimed toward the development of commercially viable industries based on space technologies that would be export-oriented and internationally competitive.

The co-ordination and management of the NSP was assigned to the Australian Space Office (ASO), established within DITAC in 1987 as the secretariat of the Australian Space Board. Its areas of responsibility included the administration of the Australian relationship with NASA, oversight of NSP-supported space projects, monitoring progress on private sector launch facility proposals and the development of national space policy. Despite the Madigan Report's recommendation of annual funding for the National Space Program in the order of $20 million per year over five years, so as to develop an effective level of space activity, the budgets allocated to the ASB were in the vicinity of $3-$5 million per year. This meant that promising projects, such as FUSE/Lyman, were terminated, due to lack of funds, while other opportunities were not taken up because the available funding was already fully committed.

Despite its financial limitations, the ASO supported the Endeavour space telescope project and, while it was criticised for funding feasibility studies for projects that it would probably not be able to financially support to completion, by the time of the first review of the National Space Program in 1992, it had contributed to the growth of the Australian space industry to the extent that there were some 24 companies or organisations claiming various degrees of space-qualified capability.

In 1990, the ASO implemented an Australian Space Industry Development Strategy, which concentrated

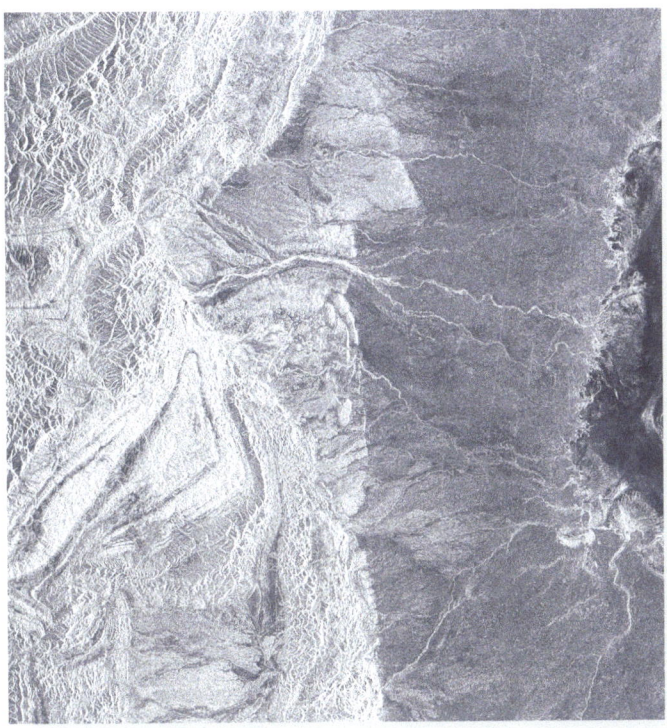

Above An image of an area near Lake Frome in South Australia, captured by the ERS 1 satellite and processed with CSIRO's AETHERS system, specifically designed for ERS images. (Image courtesy of CSIRO)

on satellite communications, remote sensing and space launch services – the last a response to the growing number of proposals for commercial spaceports and launch activity in Australia (see below). To foster commercially oriented research and development in space-related technologies, the Space Industry Development Centre (SIDC) program was initiated. SIDCs were to be jointly funded partnerships, aimed at bringing together research institutions and commercial enterprises in the development and eventual marketing of commercially viable space products and services. With funding supplied by both the National Space Program and private industry partners, SIDCs were intended to encourage space-related research and development, provide a visible focus for it, and promote industry and university collaboration in achieving it.

By 1992, four SIDCs had been approved, the most successful of which would be the Australian Centre for Signal Processing at the University of South Australia. This centre would become part of the university's Institute for Telecommunications Research, established in 1994 and still active today as one of Australia's most significant research groups in the area of wireless telecommunications. The Space Centre for Satellite Navigation at the Queensland University of Technology, with a focus on GPS applications, would also play an important role in the FedSat project.

1992 was a pivotal year for the National Space Program, which was the subject of two reviews, further discussed in Chapter 9. Both assessed the performance of the NSP from very different perspectives and arrived at markedly different conclusions. Commissioned by DITAC to assess the performance of the National Space Program and make recommendations for its future direction, the Expert Panel review, undertaken by three recognised space experts, determined that, despite certain criticisms and low budgets, the NSP and the ASB/ASO had performed well in the relatively short period of their existence in developing Australian space industry capability. However, the concurrent economic evaluation of the NSP conducted by the Bureau of Industry Economics (BIE) concluded that there was no economic case for an independent National Space Program.

Although the Keating Government eventually chose to accept the recommendations of the Expert Panel, its support for those recommendation was less than wholehearted. Noting criticisms in regard to the co-ordination of space activities, the report recommended a new, high-level Australian Space Council that would formulate space policy and set priorities for a revitalised National Space Program. While the Government did replace the ASB with a statutory Australian Space Council (ASC), supported by an enhanced Australian Space Office within DITAC, it did not substantially increase the NSP budget, as recommended by the Panel. This left the revised National Space Program, focused on remote sensing for environmental monitoring, mobile communications systems, aspects of launch services, space science and continued growth of the space industry, to operate with little more funding than previously.

This drew considerable criticism, as a constant concern within the space community in the 1990s was the possibility that access to space-based data for a wide range of services on which the nation had come to rely (remote sensing, environmental, weather and scientific satellite data), currently being supplied to Australia for free, might eventually be charged for, if Australia did not 'buy in' to future satellite programs by contributing an instrument or funds. The potential 'threat' to Australian access to space data remains a concern to this day, as the nation is still without major investments in satellite systems outside the communications and Defence spheres.

Despite the successes of the ATSR instruments and, to a lesser degree, the Endeavour space telescope, and with other projects in development, a new review of the NSP was announced in 1995. Although the report of the Interdepartmental Committee conducting this review approved of the NSP's efforts in developing national space capability, the National Space Program, the ASC and the ASO were all terminated in June 1996 by the new Howard government.

Going Up from Down Under: Spaceports and Launch Systems

Although global satellite telecommunications was based on the use of large satellites in geostationary orbit, the development of powerful computing and signal processing technologies in the 1980s made the use of LEO satellites for communications feasible and an attractive idea for the development of global mobile communications systems at a time when the terrestrial mobile cell network was in its infancy. By the late 1980s, there were several proposals for constellations of small, relatively inexpensive satellites, which could be launched reasonably cheaply into Low Earth Orbit, to provide 'switchboards in the sky' for mobile networks. Groups of satellites would occupy different positions in each of a number of different orbital planes: this would ensure that there would always be at least one satellite in the line of sight of any location on Earth.

Successful operation of planned LEO communications systems would require dozens, if not hundreds, of satellites in orbit, requiring regular replacement of the cheap individual satellites, which were not expected to have long operational lifespans. The large numbers of small satellites required for LEO satellite systems raised expectations that there would be a significant increase in launch service requirements, which

ASRI: More than Amateur Rocketry

In the early 1990s, the space industry focus of the NSP encouraged the amalgamation of two 'amateur' space associations, the AUSROC Launch Vehicle Development Group at Monash University in Melbourne and the Australian Space Engineering Research Association (ASERA), into the volunteer-run Australian Space Research Institute (ASRI).

AUSROC, which began in 1988 as a student project at Monash University in Melbourne, was a volunteer group that aimed to design, manufacture and launch small rockets as a publicly visible space education program. Its long-term goal was to progressively construct and launch larger sounding rockets, eventually developing the world's first 'amateur' satellite launch vehicle. In 1992 and 95, AUSROC launched two small AUSROC II sounding rockets at Woomera. ASERA was a volunteer-run organisation formed in 1990 to provide a means of coordinating a non-commercial space technology research and development program. Its planned projects included an amateur radio satellite conceived as a successor to Australis-OSCAR 5.

The two groups merged in March 1993, intending to embark on an ambitious plan of sounding rocket and small satellite development. While lack of funds prevent ASRI from achieving its major sounding rocket goals, from 1997 until recent years it conducted a regular program of small rocket firings at Woomera using Zuni and Sighter rockets made available by the RAAF. As part of its educational goals, ASRI offered flights to school student projects as prizes in space contests run by the Institution of Engineers, Australia and the Kistler Aerospace Corporation. ASRI members have also carried out projects with universities, space industry and international space agencies.

Completely run by volunteers, and without any government funding, ASRI has been an important training ground for Australian students interested in space engineering and other areas of space technology and many of today's Australian space professionals gained early experience working on ASRI projects. The organisation continues today as an important contributor to Australian space education and space workforce development.

would not be met by the existing launch service providers, especially as geostationary launch systems were not well suited to the requirements of LEO systems. This 'launch service gap' was seen as an opportunity for Australia to develop commercial launch facilities and/or launch vehicles, to serve what was expected to be a burgeoning market.

Woomera

With Woomera lying idle in the 1980s, plans were put forward for its revitalisation based on the redevelopment of the old ELDO facilities to provide the basis of a light satellite launch service, utilising the Range's polar orbit launch corridors to take advantage of the expected LEO launch market. The first proposal of this nature was put forward in 1988 by an Adelaide-based consortium, Australian Launch Vehicles Pty Ltd (ALV), backed by the Transfield engineering group, which intended to develop a new Australian light launcher for the service. Their proposed vehicle, the Australian Launch Vehicle (ALV), was to be a low-cost three-stage rocket combining local expertise with proven foreign technologies and hardware, capable of launching up to 1000kg. However, ALV's plans were based on securing at least part of the launch contract for the Iridium series of satellites in planning by Motorola: when the Iridium opportunity did not eventuate, the ALV group decided not to pursue the project further.

In 1991, the Southern Launch Vehicle (SLV) consortium was formed to continue where the ALV proposal left off. Composed of three Australian space industry companies, Auspace, British Aerospace Australia and Hawker de Havilland, with the backing of the South Australian state government,

ABOVE *The last vestige of the Kistler project at Woomera, the sign marking the location of the company's planned launch facility, picked up for this photo from where it had earlier fallen. (Author's collection)*

the SLV group looked to develop a light satellite launch service to complement the geostationary launch service that was proposed for the Cape York Space Port in far north Queensland, and also to utilise Woomera as a recovery area for re-entry capsules used in microgravity experiments. Intended to be a low-cost, three-stage launcher that would require minimal infrastructure and support equipment, the SLV rocket was intended to use purchased US rocket engines combined with Australian fabrication to create a simple vehicle adaptable for a range of launch services for satellites in the 750-1000kg range. The first two test launches of the SLV were planned with the intention of providing a launch opportunity for Australian satellites, to aid local space industry development. The first test launch, qualifying the key system elements of the launch vehicle and its ground control segment, would carry the RASS (Remote Area Satellite Service) satellite, a proposed small store-and-forward communications satellite for collecting data from instruments in remote locations. The second test launch, which would qualify the system for operational applications, was planned to carry an Australian remote sensing payload, such as CSIRO's GAMS (Global Atmospheric Methane Sensor).

A 1992 Senate Standing Committee report on developing satellite launching facilities in Australia and the role of government, recommended government assistance to the SLV project and consideration of the revitalisation of Woomera. The Expert Committee report, in that same year, also made a similar recommendation. However, the repeated delays to the commencement of the LEO mobile systems on which the SLV project's viability was based led to the project's demise in the mid-1990s.

In 1998, the US-based company, Kistler Aerospace, established an office at Woomera, intending to use the Range for the test program of its K1 rocket, a two-stage fully reusable vehicle that was also aimed at supporting the LEO launch market. Although Kistler designated a launch area on the Range, and broke ground there to great fanfare, the company's vehicle was never successfully developed, despite funding support from NASA, and no launches took place in Australia. Other proposals, such as the 1999 SpaceLift Australia announcement that it would launch LEO satellites from Woomera using modified Russian SS25 missiles converted to launch commercial payloads, also failed to come to fruition.

Virgin Galactic, the space tourism company founded by British entrepreneur Sir Richard Branson, briefly considered Woomera as the launch and landing site for its Spaceship Two vehicle in the mid-2000s. However, the Australian Government did not offer any incentives for Virgin Galactic to establish itself at Woomera and the company now plans to operate from Spaceport America in New Mexico, USA.

Cape York to Christmas Island: Equatorial Spaceport Proposals

Even before the proposals for the revival of Woomera, in 1986 a commercial spaceport was proposed for the Cape York area of far north Queensland, to take advantage of its geographical advantages for launching geostationary satellites, which were discussed in Chapter 3. When the Cape York spaceport concept was first proposed by Hawker de Havilland Technical Services Director Stanley Schaetzel, there was a global shortage of commercial satellite launch services, leading to proposals for commercial equatorial spaceports in various parts of the world.

Sponsored by the Queensland State Government, a 1987 study by the National Committee on Space Engineering of the Institution of Engineers, Australia suggested that a commercial operation could be established, but would require the Cape York facility to capture at least a quarter of the world commercial launch market in order to become profitable. Although a key aspect of the proposal was to serve overseas private launch vehicle companies by offering a cheap, deregulated launch facility, open to all comers, Australian Government commitment was also required as certain international space legal agreements required Commonwealth support.

The spaceport was initially conceived as a combination of major elements including an Air Force base, seaport and tourist complex in conjunction with the actual launch facility. These combined elements were seen as necessary to the viability of the project, although later attempts to develop a spaceport facility did not include them.

Two companies formed in 1987 to undertake more detailed feasibility research into the venture: the Cape York Space Agency (CYSA) and the Australian Spaceport Group (ASG). Both companies were consortia of major engineering and financial interests, supported by smaller players. ASG included US launch vehicle supplier Martin Marietta (now Lockheed Martin). The CYSA study concluded that the venture was commercially viable in the long-term without government support or reliance on military payloads or a tourist complex, but that this viability depended upon having a marketable launcher to offer to potential clients, rather than user-supplied vehicles. ASG followed a shorter timeframe, examining the near-term prospect of constructing a launch site on the Cape's west side near Weipa. Unlike CYSA, they concluded that a spaceport was not viable and terminated their interest in the project.

CYSA, however, remained convinced of the feasibility of the project if a suitable launch vehicle could be found. In 1988, it signed an agreement with the Soviet Union for the purchase of Zenit rockets, a commercial version of the highly reliable Tsyklon launcher, capable of placing a payload of up to 2200kg into geostationary orbit. The rockets would be assembled, integrated and launched from the spaceport by an Australian launch crew. CYSA expected a launch rate of five rockets per year, commencing in 1995.

BELOW Map of northern Australia and surrounding areas, showing the location of a number of proposed sites for commercial equatorial spaceports and LEO satellite launches mentioned in this chapter.

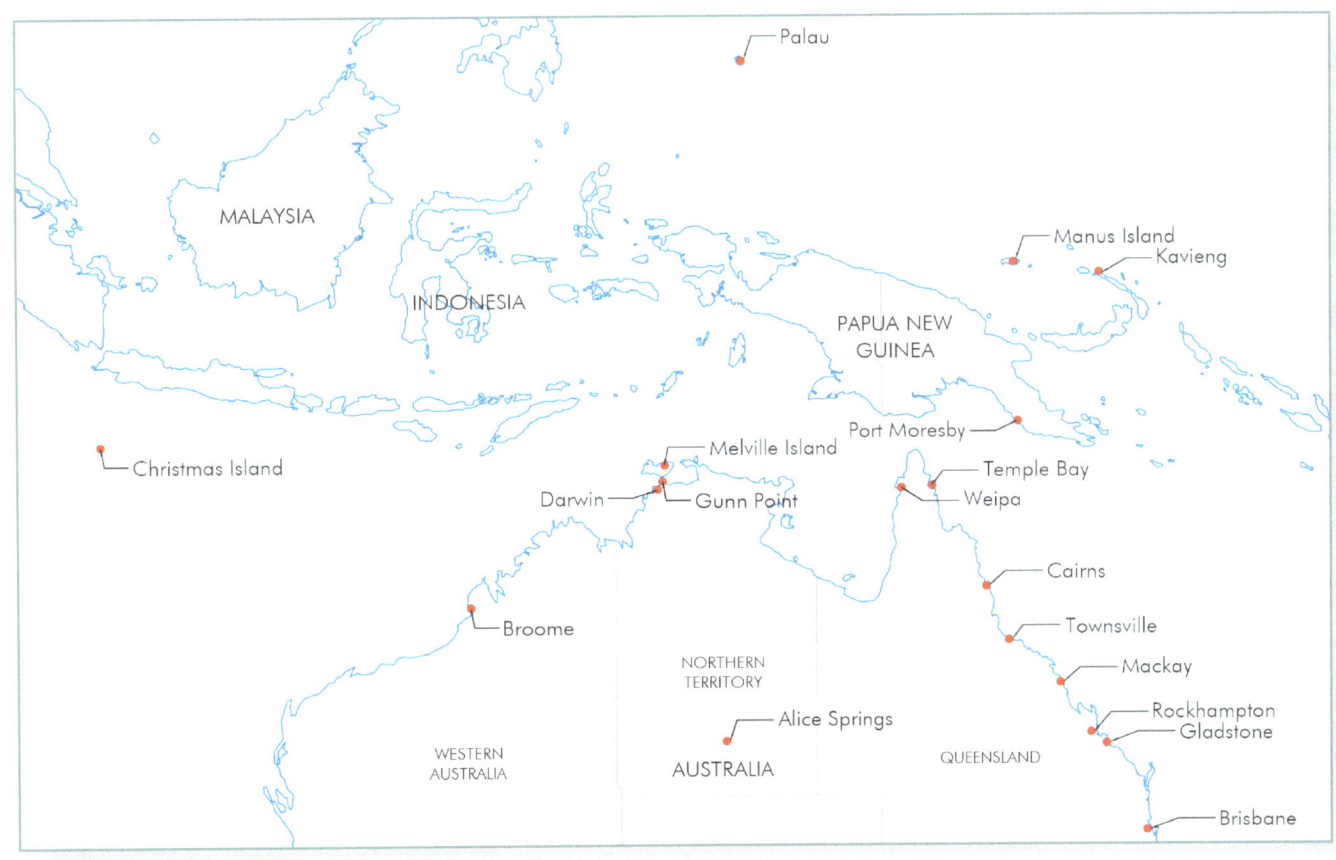

Happy Landings: Woomera as a Re-entry landing site

Woomera's vast, sparsely populated downrange was considered as a landing site for re-entry vehicles including recoverable microgravity research capsules that would return to Earth with their cargo of experiments after spending some time in orbit. The German-Japanese Express microgravity and re-entry experiment spacecraft was intended to land at Woomera after a period in orbit in 1995. However, a launch malfunction meant that the satellite was put into a very low orbit, which decayed in a few days, with the satellite re-entering to crash down in Ghana.

In June 2010, the re-entry capsule of Japan's Hayabusa asteroid exploration mission returned to Earth at Woomera carrying precious microsamples of asteroid material. The sample return capsule of the follow-on Hayabusa 2 mission, launched in 2014, is expected to land at Woomera in December 2020.

Woomera has also been considered as a landing site for the Soyuz spacecraft attached to the International Space Station as a 'lifeboat' for crews to return to Earth in an emergency. Due to the orbital path of the space station, suitable landing sites are limited (and none at all are available in the US). Consequently, if an emergency landing could not be made in Russia, Woomera could act as an alternative.

In late 1989, the Commonwealth and Queensland State Governments announced full non-financial support for the project, and in 1990, 60 000 hectares of land was purchased at Temple Bay. The Temple Bay site was not without controversy, with environmental groups expressing concern over damage to rainforest areas and the Great Barrier Reef, and Aboriginal groups concerned about the effect of the spaceport development on tribal lands and sacred sites in the area.

When CYSA experienced financial difficulties in 1990-91, the project ground to a halt and the Australian Space Office sought expressions of interest from other investors. Space Transportation Systems Ltd (STS) became the preferred developer for the spaceport, on condition that it achieved a series of developmental milestones. Unfortunately, due to the prevailing economic climate, STS was unable to meet those conditions and lost its preferred developer status in late 1992. The Australian Government then sought new expressions of interest and in 1993 another consortium, Cape York International Spacelaunch Ltd (CYISL), took over. Although the Expert Panel report recognised the potential for the nation if the Cape York Space Port could be established, it did not recommend that it be made a priority under the new National Space Program,

RIGHT *The re-entry capsule of Japan's Hayabusa asteroid mission streaks through the South Australian sky as it comes in to land on the Woomera Range. The capsule carried microsamples of asteroid material, returned to Earth for analysis. (Picture courtesy of NASA/Ed Schilling)*

since the project was viewed as a commercial, rather than a government, undertaking. CYISL, too, failed to bring the project to fruition and no further attempts were made to develop a commercial spaceport in the Cape York region.

The idea of a commercial spaceport in Australia was not dead, however. After losing its preferred developer status for Cape York, STS transferred its interest to Darwin, in the Northern Territory, proposing to use Proton vehicles provided by International Launch Services, the joint venture between Lockheed Martin and the Russian Krunichev State Research and Production Space Centre. STS investigated launching from Melville Island or Gunn Point, near Darwin, and also looked further north to possible launch sites in the vicinity of Papua New Guinea (including Manus Island and Kavieng) or in the island nation of Palau. In 1998, STS finally withdrew from attempts to develop a spaceport when ILS advised that they would not proceed with any Proton launches from Australia, due to costs. SLS was also connected to United Launch Systems, a company that, in the late 1990s, intended using the Unity launch vehicle, based on a converted former Soviet ICBM, for LEO satellite launches, based at Gladstone, Queensland. This project, too, did not proceed.

The last major proposal for a commercial equatorial launch facility was put forward by International Resource Corporation (IRC), which was backed by Asian financial interests. IRC proposed the Asia Pacific Space Centre (APSC) on the Australian Indian Ocean territory of Christmas Island, using a variant of the Russian Soyuz launch vehicle. Although the Australian Government announced support for the plan in March 1998 and funded upgrades to the airport and other facilities on the island, the project failed to meet its developmental milestones, seemingly concentrating its efforts on the establishment of a casino resort, which had been part of its overall development plan.

Although the Japan Aerospace Exploration Agency (JAXA) used the runway at Christmas Island in 2002 for three tests of its High-Speed Flight Demonstration (HSFD), a component of its HOPE spaceplane development project, no further activity occurred in relation to APSC's development and the project was considered defunct by the mid-2000s.

FedSat: Celebrating Australia's Centenary with a Satellite

With the National Space Program that had provided a mechanism for space industry development for a decade terminated, the Howard Government considered new options for assisting Australian industry to become space qualified. In August 1996, it announced a new administrative structure for Australian civil space activities, with a focus on the development of a small technology demonstrator satellite to be launched in 2001 as a celebration of the centenary of Australian Federation. Named FedSat, this new satellite would be a demonstration of Australia's intention to embrace the technologies of the Twenty First Century.

FedSat would be developed by a new Cooperative Research Centre for Satellite Systems (CRCSS), created within the existing national research structure of Cooperative Research Centres (CRC), which combined research institutions in collaboration with commercial partners to maximise the benefits of research through commercialisation and technology transfer. The CRCSS was created by transferring the core staff of COSSA, already well experienced in the development of satellite instruments through the ATSR project, to the new centre: the remaining COSSA staff were incorporated into the CSIRO's Earth Observation Centre.

In July 1997, an initial program grant of $20 million for FedSat was announced and the CRCSS commenced operation on 1 January 1998, with partners including Auspace, Optus, DSTO, Vipac Engineers and Scientists and several universities. The Space Industry Development Centre for Satellite Navigation at the Queensland University of Technology and the Institute for Telecommunications Research at the University of South Australia undertook teaching and research for FedSat's development, as part of the educational role required of CRCs.

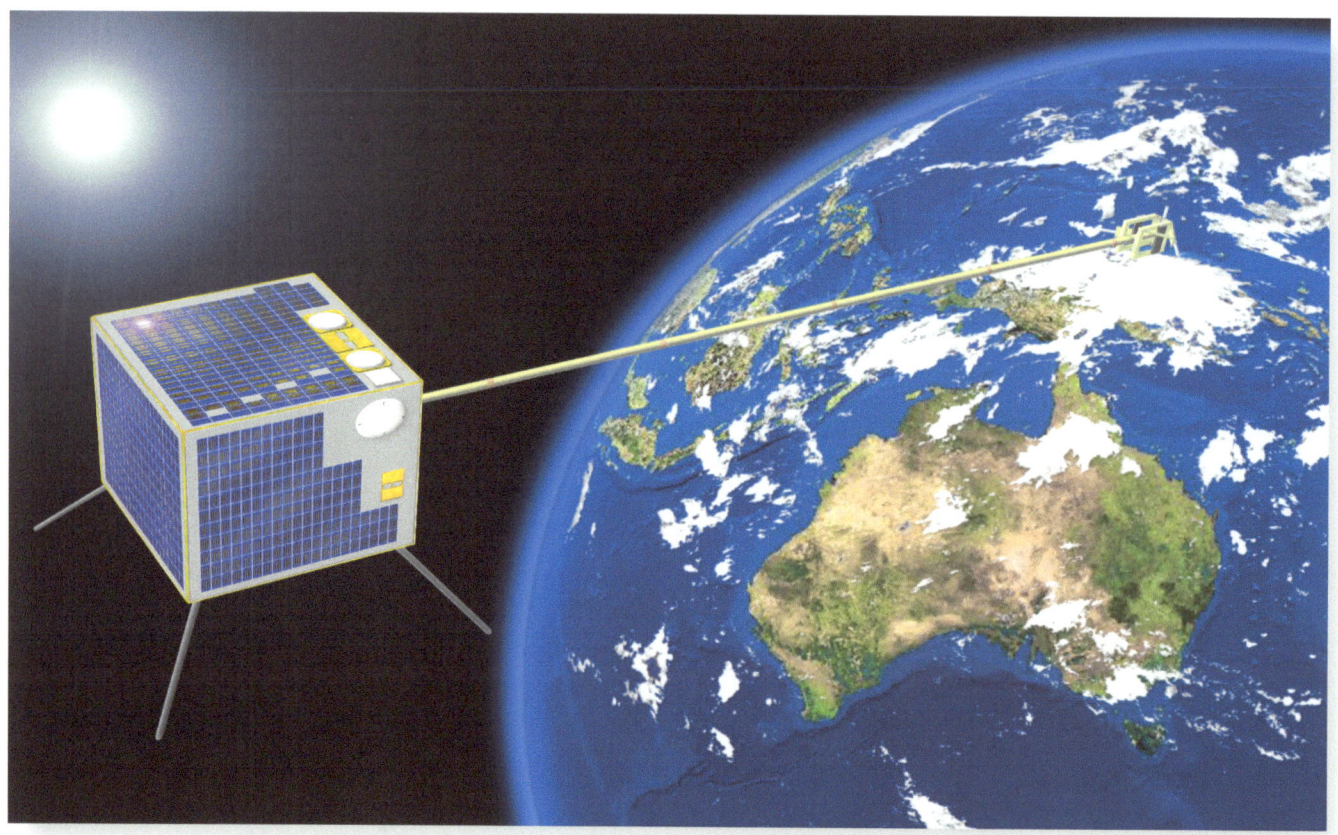

ABOVE The first Australian satellite since WRESAT, FedSat was developed by the CRCSS and partially constructed by Auspace. It carried a mix of Australian and internationally supplied experiments. Although it was not launched until 2002, it was a successful celebration project for the Centenary of Australian Federation. (Illustration by Glen Nagle)

FedSat was a 58kg microsatellite that would carry six payloads provided by Australian and international partners. The largest was the communications experiment, developed by CSIRO and the Institute for Telecommunications Research, consisting of three components (UHF transponder, Ka-band transponder and base-band processor) that were used to perform several experiments in satellite communications. The provision of the Ka-band payload made FedSat the world's first microsatellite capable of operating in that band. The High Performance Computing Experiment used a reconfigurable computer based on a Xilinx FPGA, another first of its kind in space use.

NASA provided a GPS payload for ionospheric studies between the orbits of the GPS satellite constellation, at 20000km, and FedSat's 800km orbit. The University of Newcastle (NSW) provided the NewMag payload, a magnetometer used to take measurements of the Earth's magnetic field near its poles, while the University of Stellenbosch, in South Africa, provided an experimental star camera, used to determine the position of the satellite in space by analysing images of surrounding stars. The final FedSat payload was a 'cultural time capsule' on CD, which included songs and recorded statements from several hundred Australian school children on their vision of a future Australia and its place in the Asia–Pacific region.

Due to the short timespan for the satellite's development to achieve a launch in 2001, it was decided to purchase the satellite platform from UK company Space Innovations Limited (SIL). Unfortunately, SIL went into receivership while the satellite was under construction, causing a delay to the program while the partially-completed satellite was retrieved from the UK and completed by Auspace, which also managed the integration of payloads at its Canberra facility.

FedSat's launch was provided free as a centenary gift from Japan, although, due to the construction delays, the satellite was not launched until 14 December 2002. Soaring into space as Australia's first satellite since WRESAT, FedSat was the first foreign satellite to be launched by Japan's H-IIA vehicle. Overseen by the CRCSS, FedSat operated for the next three years, testing new technologies in satellite computing, positioning technologies and communications, as well as gathering data on space weather and radiowave propagation.

The CRCSS was intended to have a broader role than just the development of FedSat, undertaking research in space science, satellite communications and satellite systems and engineering, aimed at developing significant Australian capabilities in satellite technologies. The Centre provided data and messaging systems to two foreign satellites and made the world's first demonstration of self-healing space computers, in addition to attracting important grant funding and more than $4 million in-kind contributions from external research collaborators. Despite its modest success, funding for the CRCSS was not renewed in 2005 and it closed at the end of that year. The still operational FedSat was transferred to the control of the Department of Defence, which operated the satellite until its battery failed in 2007, bringing the project to an end.

FedSat was also to be the last significant space project for Auspace, which tried unsuccessfully from 1998 to attract funding to its ARIES satellite project. The Australian Resource Information and Environment Satellite was a joint project of Auspace, CSIRO and ACRES for a hyperspectral remote sensing satellite for geological exploration

The NSSA: Public Support for Australian Space Industry

Public 'pro-space' groups, created to enable space interested members of the general community to share their interest and support for space activities, have existed in Australia since the late 1950s, when the first local chapter of the British Interplanetary Society was established. Once group stands out for its professional level of support for the development of Australian space industry and the establishment of an Australian space program and space agency – the National Space Society of Australia (NSSA).

An off shoot of the US-based National Space Society, the NSSA was established in the 1980s. Driven by national President Kirby Ikin, himself involved in the space re-insurance business, the Society adopted a dual approach to supporting Australian space activities, with both 'grass roots' local chapters around the country, offering a local focus for space-interested individuals, and the operation of a professional level program which included the organisation of the Australian Space Development Conferences (ASDC).

First held in 1990, the Australian Space Development Conferences rapidly became an important forum for bringing Australian and Asian space companies and civil programs together with international space industry participants, especially during the downturn in Australian space activities across the first decade of the Twenty First Century. The ASDCs have offered the opportunity for Australian and regional space companies to showcase their space-related skills and achievements to all levels of government, as well as to the general public through generating strong media interest. The NSSA was also instrumental in the formation of the Australian Space Industry Chamber of Commerce (ASICC) in 1992, now the Space Industry Association of Australia (SIAA), which provides a voice for Australian space industry and works to promote its growth.

Since 2000, the NSSA has also organised the annual Australian Space Science Conferences, as an opportunity for university researchers, scientists and engineers to come together to discuss space related science and technology in Australia. Renamed in 2014 as the Australian Space Research Conference, to reflect a broadening focus on space-related research in fields beyond 'space science', such as education and heritage, this annual gathering has become an important forum for providing the space-related science and university sector an opportunity to raise issues of importance to its community.

Educating the Next Space Industry Workforce

To maintain, or grow, the Australian space industry sector requires a regular flow of suitably trained scientists, engineers and technologists entering the workforce. To encourage young people into the university studies that will lead to space industry careers, motivating then to undertake STEM subjects at school is an important step. The fascination that many young people have for astronomy and space exploration has long been recognised as a way of inspiring students into selecting STEM subjects in their high school studies.

From 1965 until the early 1970s, NASA operated a program in Australia called the Spacemobile. This educational van was operated by a team of teachers who visited schools and universities in the eastern states, giving presentations to classes, as well as offering public outreach programs. With the creation of the first Space Camp in the United States in 1982, Australian schools began to offer international excursions to the US, including visits to Space Camp and NASA facilities as a way to encourage student interest in space and STEM subjects. Many of these programs have been the initiatives of individual teachers and a number of schools, and independent space education programs like One Giant Leap, continue today to offer Space Camp tours to the US or local 'space camp'-style education programs driven by passionate teachers dedicated to sharing their love of space and their belief in the value of STEM education. NewSpace company Gilmore Space Technologies has even invested in its own Space Camp-type facility in Queensland.

The first independent space education program to have a significant impact on encouraging STEM-interested high school students into space studies was the Australian International Space School, which operated for two decades as a school holiday program from 1990. Its educational director, Dr. Jeanette Rothapfel Dixon, has become a leader in training teachers to use space in the classroom, as a way to excite interest in STEM subjects. Many Space School alumni are now involved in Australian and international space activities and a South Australian offshoot from the original program is still in operation.

The first space education program to be fully operated by a State Education Department was the Victorian Space Science Education Centre (VSSEC), one of six Specialist Science Centres established by the Victorian State Government in 2006. Located in a dedicated facility at Strathmore Secondary College in Melbourne, and working in conjunction with its science teachers, VSSEC uses the context of space to engage teachers and students in the teaching and learning of STEM subjects and is now a leader in space-focussed school education in Australia.

Through immersive hands-on programs in its simulated Martian surface and Research Laboratory areas, in addition to other curriculum-linked programs, VSSEC has engaged the passion for space in many students who are now studying science and engineering at university, or embarking on careers in space science and industry. Many of VSSEC's programs can also be accessed online by schools around Australia, giving it national reach beyond its home state.

While VSSEC's programs are mainly focussed on secondary and primary school learning, it also encourages excellence in university level science and engineering studies through the VSSEC-NASA Australian Space Prize, which is open to fourth year undergraduate and honours students. The competition offers the opportunity for an Australian student to participate either in a NASA Academy program in the US, or a NASA Summer Internship. VSSEC also international reach, supporting space education programs at the annual International Astronautical Congress.

Above The 'Wombat XL' Space Simulation Facility at the Advanced Instrumentation and Technology Centre in Canberra. It is the only facility of its kind in the country, capable of accommodating both large astronomical instruments and small spacecraft. (Photo courtesy of AITC, ANU Research School of Astronomy and Astrophysics)

and mapping. ARIES would have the ability to map the mineral composition of surface rock, dirt, sand and water in 100 different frequencies. Despite its potential to reduce ground survey costs for the mining industry, the satellite was unable to secure development support and following the conclusion of its work on FedSat, Auspace struggled to find contracts in an environment where the Government was no longer actively supporting space industry development. In 2007, Auspace was acquired by the Nova Group of companies, which turned it toward the communications field.

The Advanced Instrumentation and Technology Centre: a New Facility for Australian Space Industry

Following the complete destruction of the Mount Stromlo Observatory in the catastrophic Canberra firestorm of January 2003, the Research School of Astronomy and Astrophysics at ANU made a forward-looking decision to develop the Advanced Instrumentation and Technology Centre (AITC) to support Australian astronomy and space science. Completed in 2006, the first stage of the AITC placed Australia at the forefront of astronomical instrumentation development, enabling the country to join the Giant Magellan Telescope project.

When the Government made an additional investment in the AITC in 2009 to expand its capabilities, the technologically similar requirements of astronomical and space-based systems presented an opportunity to expand the national capability in astronomical instrument manufacturing to support local space industry. New facilities were added to the Centre for the assembly, integration and testing of space systems and small satellites, including Australia's only Space Simulation Facility. This 3m diameter, 4m long thermal vacuum chamber can both simulate the airless space environment and also subject a test article to its extreme temperatures. Other new equipment included a computer-controlled 'shaker' table, capable of exerting acceleration forces equivalent to those of a rocket launch, ensuring that a satellite or instrument can survive the rigours of launch, along with clean rooms, workshops and a RF anechoic chamber.

The provision of these new facilities coincided with the implementation of the Australian Space Research Program (ASRP), enabling the AITC to support five ASRP projects and their ongoing development. The AITC provides an important focus for space industry development, in place now to support any new Government or commercial initiative to further Australia's engagement with space activities.

THE AUSTRALIAN SPACE RESEARCH PROGRAM

The 2008 report from the Senate Standing Committee on Economics, *Lost in Space? Setting a New Direction for Australia's Space Science and Industry Sector*, which is more extensively discussed in Chapter 9, included a recommendation that the Australian Government assume a greater role in co-ordinating and leading the development of space policy and the space sector. While not taking up all the recommendations of this report, the Rudd Government initiated a limited re-engagement with civil space activities, instituting the Australian Space Research Program (ASRP) in 2009.

The objective of the ASRP was to develop Australia's niche space capabilities, through support for space-related research, innovation and skills in areas of national significance or excellence. $40 million was allocated to fund a series of three-year project grants in two 'streams': space science and innovation projects, to support collaborative projects for niche space capability development; and space education development projects to support student, or student-focused, projects and space education activities. The ASRP funding was allocated to 14 projects over four rounds of funding (four in education and ten in space science and innovation), achieving direct co-investment from industry of $39.1 million and establishing a number of ongoing projects.

The four projects selected for education stream grants were:
Pathways to Space: Empowering the Internet Generation, a collaboration between the University of NSW, the University of Sydney and the Powerhouse Museum that sought to address skills shortages in Australia's space industry, through the delivery of an interactive educational program aimed at Year 10-12 students. This project has created the Mars Lab at the Powerhouse Museum, Sydney, which now offers a suite of interactive education programs for high school students based around astrobiology and the exploration of Mars, drawing national and international schools through its internet-enabled programs. The project's simulated Martian surface, the 'Mars Yard', also provides a testbed for space robotics research on roving vehicles, undertaken by the Australian Centre for Field Robotics at the University of Sydney.

Southern Hemisphere Summer Space Program, a collaboration between the University of South Australia and the International Space University to develop an intensive, five-week program of interdisciplinary space studies for professionals, graduate researchers and senior undergraduate students. Now an annual offering at the University of South Australia, this course draws students not just from Australia, but across the globe.

Place and Space: Perspective in Earth Observations, a pilot program from Flinders University to create science teacher champions who would build capability and capacity through linking inquiry-based learning with university science and engineering research.

Warrawal, a Comprehensive Tertiary Education Program in Satellite Systems Engineering, a project from the University of NSW to develop and deliver a two-year Master's degree in satellite systems engineering. This course, the first of its kind in Australia, commenced operation in 2013.

The ten projects of the space science and innovation stream were:

Platform Technologies for Space, Atmosphere and Climate, a project from the Royal Melbourne Institute of Technology, focused on advancing space-related technologies that address key research problems in climate change, severe weather events, space weather and orbital debris collision avoidance and mitigation.

SCRAMJET-based Access to Space Systems, the initial phase of the University of Queensland's SCRAMSPACE project discussed in Chapter 7.

Antarctic Broadband: Definition and Capability Development, a project led by Aerospace Concepts Pty Ltd to develop a high-quality broadband communications service for Antarctica, using small-satellite technology.

Garada-Synthetic Aperture Radar Formation Flying, a UNSW project to develop a technology for using formations of small synthetic aperture radar satellites that could collect environmental information, such as soil moisture, and assist with disaster management.

Automated Laser Tracking of Space Debris, a space situational awareness project from EOS Space Systems Pty Ltd, aimed at demonstrating how a fully automatic, remotely operated laser-based tracking station could significantly improve the accuracy of satellite orbit prediction.

The GRACE follow-on mission, an ANU project to developed hardware for a laser ranging system suitable to be flown on NASA's GRACE Follow-on mission.

Unlocking the Landsat Archive for Future Challenges, a project led by Lockheed Martin Australia to enable the ongoing processing of the national Landsat archive, which contains valuable remote sensing data from 1972 to the present, allowing studies of environmental change over time.

Space-based National Wireless Sensor Network, a project by the University of South Australia to develop a cost-effective means of data retrieval from large numbers of remotely located sensors and devices.

The Australian Plasma Thruster, which aimed to develop the HDLT plasma thruster technology invented at ANU into a flight-ready thruster design.

Greenhouse Gas Monitor, a project from Vipac Engineers & Scientists that sought to develop a sensor for measuring greenhouse gases nationally and globally.

Several of these projects have produced significant outcomes. The GPS receiver used in the Garada formation flying project was adapted for the DST Group's Biarri cubesat project outlined in Chapter 5, while another GPS receiver developed by UNSW is flying on its EC0 cubesat, launched in 2017 (see below). The Automated Laser Tracking project has led to a contract for a laser ranging telescope with adaptive optics to be undertaken by ANU and EOS Space Systems, with the Korea Astronomy and Space Science Institute (KASI). Technologies developed during the GRACE Follow-on Mission project led to the founding of a new company, Liquid Instruments Pty Ltd, that now has commercial electronic devices derived from its technology available in Australia and overseas.

The impetus given to improved government understanding of the threat to the world's orbital infrastructure from space debris by the Automated Laser Tracking of Space Debris project contributed to the formation, in 2014, of the Space Environment Research Centre (SERC), a CRC based at Mount Stromlo Observatory. A collaboration between ANU, the Royal Melbourne Institute of Technology, EOS Space Systems, Optus and international partners NASA Ames Research Centre, Lockheed Martin and the National Institute of Information and Communications Technology from Japan, SERC has been created to focus research on tracking space debris, improving predictions of space debris orbits and predicting and monitoring potential collisions in space. Building on the expertise of EOS Space Systems, it will also develop ways to modify the orbits of space debris to help avert collisions.

Although funding for the ASRP was not renewed in 2013, and the program was subsequently terminated by the incoming Abbott Government in 2014, an independent assessment of the program in 2015 concluded that it had achieved its objectives in developing niche space capabilities that offer future commercial benefits

From SCRAMSPACE to SPARTAN: Hypersonics Takes Off

Despite the launch failure of the SCRAMSPACE project, to which the ASRP grant contributed, the Centre for Hypersonics at the University of Queensland is moving forward on a new hypersonic research project, dubbed SPARTAN (SCRAMJET Powered Accelerator for Reusable Technology AdvaNcement), for the development of a partly re-usable three-stage satellite launcher that would use a SCRAMJET as its second stage, with the first and third stages rocket-powered.

The first stage booster, called the Austral Launch Vehicle, is being developed by Heliaq Advanced Engineering. Using deployable wings and a small propeller, the rocket will fly back to base. A successful first test of this system was carried out at Roma, Queensland, in December 2015. The SPARTAN SCRAMJET second stage will also fly back to Earth for a runway landing. Only the third stage will be not re-usable, burning up on re-entry after releasing its payload into orbit.

The long-term goal of the SPARTAN project is to have the operational satellite launcher flying from a launch site on the north coast of Queensland.

---

and public good to the nation, that the program had been delivered efficiently and that it had represented a value-for-money investment of Commonwealth funds.

NewSpace in Australia: a New Kind of Space Industry

NewSpace, also known as 'Space 2.0', is a new paradigm in space activity that builds upon the digital and technological revolutions of the past decade to focus on exploiting the relatively low-cost capabilities of smaller spacecraft and cheaper launch technologies. NewSpace projects are based on the rapid development of new capabilities that accept much lower technology thresholds (and consequently higher risks) than the traditional space sector, resulting in significantly lower start-up costs and faster time to market. The NewSpace industry is entrepreneurial, frequently targeting new niche markets with simple technological solutions, or opening up new dimensions in existing markets that were not cost-effective for traditional space operators to exploit.

Space 2.0 opens new possibilities and opportunities for Australian space industry, since its lower capital costs and less demanding technology requirements enable even small start-ups with a unique technology, product or service idea to directly compete in the global marketplace. The constantly-increasing capabilities and relatively low cost of small satellites, particularly cubesats, not only improve access to space for the commercial sector, but also allow research and education institutions with limited budgets to fly experimental payloads more regularly. National needs for specific space-based data and services can be more effectively met by the development of small satellite constellations carrying instruments specifically tailored for Australia's unique requirements, and at relatively low cost to Government – thus freeing Australia from its reliance on space-based data provided by other countries. Many of the technologies developed and tested under the ASRP have applications that can be exploited using a NewSpace approach.

A new generation of young Australians has embraced the NewSpace philosophy and the opportunities it offers to participate in space activities. To give just a few examples: the Silicon Valley start-up Planet Labs, that provides constant imagery of the Earth from orbit using a constellation of over 150 small satellites (making it the manager of the largest satellite fleet in the world), was co-founded by Australian Chris Boshuizen, who had previously worked for NASA; Saber Astronautics is a Sydney-based space engineering research and development company with a global clientele for its advanced satellite mission control software and satellite operations services; Adelaide-based Fleet Space Systems, a 'next-generation connectivity company' aims to connect the world and its growing Internet of Things through cutting edge communications delivered via space technologies; Gilmour Space

Technologies, based in Queensland, is combining composite materials and 3D printing technology to produce low-cost sounding rockets and orbital launch systems.

The NewSpace philosophy has also been harnessed for STEM education and space workforce training. Australian company Cuberider has developed an interactive STEM education program for schools based around cubesat technology. It offers students the opportunity to fly their experiments to the International Space Station for on-orbit operations – in fact, in 2016, Cuberider's experiments were the first ever Australian payload on the ISS. UNSW's BLUEsat undergraduate student group has projects based around the development of a stratospheric balloon vehicle for scientific experiments, the development of a cubesat and the construction of a robotic 'Mars rover'. The Melbourne Space Program, based at the University of Melbourne, is bringing together students, academics and alumni to work on a range of science and engineering-based projects to enhance professional development and gain experience in space technologies. This program is currently developing a cubesat for launch in 2018.

The story of Australian space industry is one of boom and bust, waxing and waning as successive Australian governments have embraced and retreated from space activity, providing, and then withdrawing, industry development funding. Yet, despite the frustrations of lost opportunities and investments at times seemingly wasted, Australian companies and research institutions have been fired by the challenge and excitement of participating in space activities and repeatedly demonstrated that they can produce world-leading ideas, innovations and technologies that have the potential to become valuable commercial products. They have also been at the forefront of seeking to encourage the Australian Government to develop a formal space policy and a national space agency.

The NewSpace revolution, with its lower costs of entry into the market, requiring less reliance on Government funding initiatives, holds out the promise that Australian space industry may soon, after many decades of struggle, become a larger player in the global space industry, realising the economic, technological and public good benefits to the nation that a strong space industry can provide.

The Southern Hemisphere Space Studies Program

One of the most important workforce education activities in Australia today is the Southern Hemisphere Space Studies Program (SHSSP), which is arguably the most successful project initially funded under the ASRP's education stream.

Operating since 2011, the SHSSP is an intensive, five-week, live-in program hosted by the University of South Australia that provides a unique environment for learning, networking and developing intercultural understanding. Developed in conjunction with the International Space University (ISU), the SHSSP is based upon its international, intercultural and interdisciplinary educational philosophy, which provides a well-rounded overview of the key principles and concepts in space science, space applications and services, space systems engineering and technology, space policy and legal issues and space business and project management.

Offered as a program for graduate researchers, undergraduate students in the final two years of their studies and professionals and managers in industry, government and the Defence services, more than 200 participants have taken part in the program since its inception. Some forty-five percent of program participants have been Australian, with others coming, not only from other Southern Hemisphere nations, but also from major spacefaring countries of the Northern Hemisphere. Many of its graduates have already found employment in local and international space projects and are active in NewSpace companies.

An Australian Space Trio: QB50 Satellites in Orbit

On 25 and 26 May 2017, three Australian cuebsats were launched into orbit from the International Space Station, joining Biarri Point as the first Australian satellites in orbit in 15 years. These science and technology demonstrators are part of the international QB50 project. Managed by the von Karman Institute for Fluid Dynamics in Belgium, QB50 is a network of 50 cubesats placed into 320km altitude orbits to carry out multi-point, in-situ observations of the lower thermosphere and ionosphere, as well as re-entry experiments. Each two-unit cubesat (10cm x 10cm x 20cm) carries instruments provided by the von Karman Institute, in order to obtain consistent data from all satellites, in addition to experiments provided by each individual satellite contributor. As their orbits decay, the cubesats will be able to explore progressively lower and lower layers of the ionosphere and thermosphere that have been very little studied to date. The mission lifetime of individual Cubesats is estimated to be between three and nine months.

UNSW EC0

Commenced in 2012, the UNSW-Educational Cubesat Zero (EC0) QB50 project has been led by the Australian Centre for Space Engineering and Research (ACSER), in collaboration with CSIRO's Data61. Founded in 2010, ACSER has research strengths in areas including Global Navigation Satellite Systems receiver design, remote sensing satellite systems and novel satellite structures utilising rapid manufacture: it is also an emerging leader in research into Off-Earth Mining technologies. Data61 is Australia's leading data innovation group, which was formed in 2016 from the integration of CSIRO's Digital Productivity flagship and National ICT Australia Ltd (NICTA).

In addition to the instruments supplied by the von Karman Institute, UNSW-EC0 carries four experiments:

- UNSW Kea Space GPS – This experiment will 'space qualify' the new Kea GPS board, demonstrating its ability to provide the satellite's position and velocity in orbit. It will also be used to carry out ocean remote sensing using GPS-reflectometry
- NICTA seL4 Computer – Intended to demonstrate the use of the seL4 operating system in critical system operation in the space environment. The experiment will also monitor the fault tolerance of the system in a radiation environment.
- UNSW RAMSES (Rapid Manufacture of Space Exposed Structures) – The body of EC0 is a rapid prototyped 3D printed structure and this experiment will demonstrate the use of this type of structure in the space environment.
- RUSH (Rapid Recovery from SEUs in Reconfigurable Hardware) – The primary objective of this experiment is to demonstrate and validate new approaches to rapidly recovering from Single Event Upsets (SEUs) in reconfigurable hardware. A SEU is a change of state caused by an ionising particle striking a sensitive node in a micro-electronic device. Secondary objectives of this experiment are to map SEU event occurrences in the thermosphere and to demonstrate in-orbit reconfiguration.

INSPIRE 2

INSPIRE 2 (Integrated Spectrograph Imaging and Radiation Explorer) is collaboration between the University of Sydney, the University of NSW and the Australian National University, with a focus on scientific research. Led by the Sydney University School of Physics and the School of Aerospace, Mechanical and Mechatronic Engineering, INSPIRE 2 uses commercial 'off-the-shelf' components for its basic structure and control functions, and carries four instruments developed by the University of Sydney and UNSW:

- Nanophotonic Spectrograph (Nanospec) – Nanospec is a novel nanophotonic spectrograph in the visible range that has been developed by the Space Photonics Group and the Institute of Photonics and Optical Science at the School of Physics, University of Sydney.
- Radiation Counter – The Radiation Counter is a standard Geiger tube system that is very low mass and relatively small. Developed by the Fusion Plasma Physics Group at the School of Physics, the Radiation Counter will count high energy photons (X-rays and gamma rays) in the space regions through which it passes.
- Microdosimeter – A standard solid-state scintillator system that is very low mass and uses little power, the microdosimeter was developed in the Space Engineering Group, School of Aerospace, Mechanical, and Mechatronic Engineering.
- Kea GPS Receiver – The Kea GPS receiver is a payload developed by UNSW of the same type as that installed on the UNSW-EC0 cubesat.

INSPIRE 2 was assembled in the Faculty of Science at the University of Sydney and at ACSER. Both ACSER and the University of Sydney (represented by its Sydney SpaceNet collaboration) are founding members of the Delta-V Space Alliance, Australia's space startup accelerator. Established in 2014, Delta-V aims to grow Australia's NewSpace commercial sector by bringing together startup companies, customers and investors. Other members include Saber Astronautics, Hypercubes Pty Ltd and Cuberider.

LEFT ACSER's EC0 cubesat, developed for the QB50 project, in the laboratory at the University of NSW. (Photo courtesy of the Australian Centre for Space Engineering Research, UNSW)

On 5 June 2017, the Australian Government announced a grant of more than $4.5 million from the Australian Research Council to the University of Sydney to establish a Training Centre for CubeSats, Unmanned Aerial Vehicles and Their Applications. In partnership with the University of NSW and with 11 other industry, government laboratory and international university partners, this new Centre aims to train the next generation of workers in cutting-edge advanced manufacturing, entrepreneurship and commercial space and UAV applications.

SUSat

SUSat is a collaboration between the University of Adelaide and the University of South Australia. Designed to focus on upper atmosphere science and radio communication experiments, the SUSat project has also incorporated education, training and outreach goals. Constructed at the University of Adelaide, SUSat carries an advanced communications payload from the University of South Australia's Institute for Telecommunications Research, in addition to a von Karman Institute supplied Ion/Neutral Mass Spectrometer.

At the time of writing, both EC0 and INSPIRE have been confirmed as successfully operating in orbit, although the status of SuSat is unknown.

BELOW The QB50 cubesats undergoing testing in the AITC's Space Simulation Facility. EC0 is on the left, INSPIRE 2 is at the rear of the test table with SUSAT on the front right. (Photo courtesy of Ann Cairns)

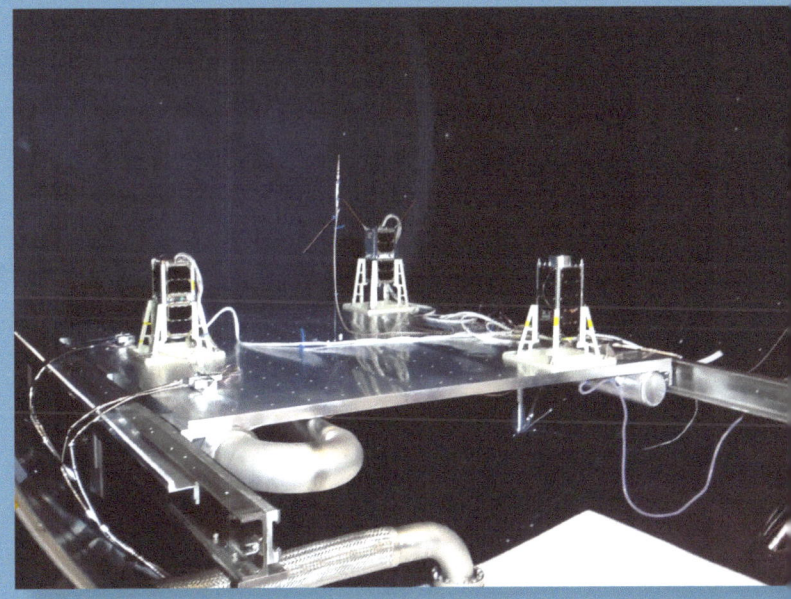

ABOVE The logo of the Australian Space Office, which was established in 1987 to co-ordinate and manage Australia's first National Space Program. (Courtesy of Department of Industry, Innovation and Science)

RIGHT The first space policy formally endorsed by an Australian Government was Australia's Satellite Utilisation Policy, released by the Gillard Government in April 2013. (Courtesy of Department of Industry, Innovation and Science)

CHAPTER 9

Ground Control: Australian Space Policy

As previous chapters have shown, Australia's space history is one of surprising achievements and disappointing lost opportunities, both situations stemming directly from Australian Government space policy (or the lack of it), since the major proportion of this country's space activities has been dependent upon Government for much of their support and funding. The fortunes of local space science and industry have been essentially dictated by Government attitudes and policy toward space activity.

Australian participation in space activities was initiated without a national commitment to an Australian space program, evolving instead from a combination of other national requirements that such activities could satisfy: defence and national security needs, support for international alliance partners, the desire to encourage the immigration of skilled scientific and technical personnel, economic development (it was estimated that every $1 Australia spent on Woomera brought in another $1 from overseas), the availability of international collaborations and Cold War geopolitical prestige considerations.

BACKGROUND IMAGE *Australian Prime Minister RG Menzies was in power when Australian space activities commenced. Neither his government nor its successors saw a need for an Australian space agency. (Courtesy of www.honeysucklecreek.net)*

Successive Australian governments until the 1980s saw no need to develop a national space policy, or to support an active national space effort, preferring instead to allow various government departments to exercise control over different programs and budgets for space activities. Despite attempts since then to initiate a comprehensive national space program, supported by a national space policy and a dedicated civil space budget, the status quo of fragmented management and budget for national space activities has prevailed to the present day.

At the Dawn of the Space Age: Space Programs Support Defence Requirements

Although the Woomera Rocket Range became the hub of Australia's first space activities, its origins, as outlined in Chapter 2, lay a decade before the dawn of the Space Age with the Anglo-Australian Joint Project and the British requirement for a long-range weapons test facility. While Australia embraced and supported the civilian scientific goals of the International Geophysical Year, its participation in IGY space activities served both civil and military purposes. The Australian sounding rocket program provided data that had not only scientific value but also defence applications, while the original Australian agreement to host the United States' IGY satellite tracking stations was accepted at least partly because the WRE recognised that the Minitrack and Baker-Nunn facilities could track military as well as civilian satellites and that both stations could be utilised as additional tracking instrumentation for the Black Knight and Blue Streak projects due to commence at Woomera soon after.

Early attempts to initiate purely civil, scientific space activities did not meet with much success. A 1959 proposal for an Australian civil space program, put forward by the Australian National Committee for Space Research (ANCOSPAR) of the Australian Academy of Science, was rejected by the government on the basis of cost, as was a similar proposal put forward by the WRE in 1968, in hopes of creating a modest Australian civil and defence space program, building on the success of WRESAT. In 1961, an offer from the US to launch a satellite for Australia (as they were soon to do for Britain and Canada) was quashed before even being considered by Cabinet, for reasons that are not readily apparent, despite strong support from the CSIRO and other sections of the Australian scientific community. Perhaps the most successful civilian space science programs of this early period were those of the Universities of Adelaide and Tasmania, which gained acceptance

Above NASA tracking stations were a facet of the close political ties between Australia and the US during the 1960s. Prime Minister Menzies recognised this and officiated at the opening of the Tidbinbilla DSN station in 1965. (Photo courtesy of www.honeysucklecreek.net)

into the UK's Skylark scientific sounding rocket program at Woomera as 'British' universities. As outlined in Chapter 3, Australia declined to become involved in the initial British proposal for a Commonwealth satellite launch vehicle, before agreeing to participate in the ELDO program, on the basis that it would be considered a full member (without financial contribution) by providing the operation and management of the Blue Streak facilities already constructed at Woomera. Consequently, while Australia was a member of ELDO, with a significant level of involvement, its role was that of a supplier of services rather than a contributing participant. As a result, launcher development activities at Woomera did not survive the transfer of ELDO operations to French Guiana. Similarly, the WRE sounding rocket programs, without a clear defence research purpose, did not survive the wind-down of the Joint Project: Australian space science was severely curtailed without access to the 'free ride' provided by the WRE and the shutdown of the Skylark program at Woomera in 1979.

With much of Australia's early space activities centred on the Woomera Rocket Range, they were overseen by the Department of Defence and the Department of Supply, which managed the activities of the WRE and the NASA space tracking stations in Australia. Day-to-day management was provided by the WRE, which, while never officially designated as such, effectively acted as Australia's 'space agency' for almost two decades. The efficient and successful manner in which the WRE co-ordinated space projects at Woomera may perhaps have been to some extent invisible to Government, lying as it did underneath the Departments of Defence and Supply. Coupled with the fact that the developing space applications of satellite communications, weather monitoring and remote sensing fell under the purview of other government departments, this 'invisibility' may have contributed to the development of the Government view that Australia had no need of a space agency and that its military and civil space activities could be managed by government departments and funded through their budgets.

The 1970s: A Low Point in Space Engagement

As both British and Australian weapons programs at Woomera began to decline, in the wind down to the closure of the Joint Project in 1980, the 1970s became a low period for Australian engagement with space. The cessation of launcher development, sounding rocket and defence hypersonic research programs, and the restructuring of the ADSS into the Defence Science and Technology Organisation, meant that space expertise and capabilities built up since the 1950s were lost as engineers and

ABOVE *The Australian sounding rocket program served both scientific and defence research needs, but without an Australian space program it did not survive the end of the Anglo-Australian Joint Project. (Photo courtesy of Defence Science and Technology Group)*

ABOVE Launch of the European Space Agency's first Ariane rocket in 1979. Although Ariane vehicles have become the world's most-used satellite launcher, Australia has repeatedly declined invitations to join ESA, passing up a potential opportunity to grow the Australian space industry. (Photo courtesy of ESA)

scientists, unable to find space-related work in Australia, went overseas or turned their attention to non-space interests.

With the demise of the ADSS, the close interrelationship between defence and civil space activities, which had begun in 1957, ended. While Australia's utilisation of space for defence and national security purposes continued to be determined by its strategic alliances, defence and foreign affairs policies, the civil space sector struggled to find government support that would enable the development of a successful national space industry serving scientific, commercial and public good space needs.

As early as 1970 an attempt was made to develop a 'post-Joint Project' civil space program, when a number of business executives, scientific bodies and industrial organisations submitted a paper to Prime Minister John Gorton, calling for the creation of an Australian Space Research Agency (ASRA). The ASRA proposal, more broadly-based than the earlier ANCOSPAR and WRE proposals, recommended the establishment of a central authority to co-ordinate and plan Australia's efforts in space research and technology, develop and commercialise the space sciences, seek overseas contracts and develop an Australian technology base.

With its emphasis on technology development and the creation of an Australian space industry that could compete for contracts in the global space arena, the ASRA proposal prefigured subsequent attempts at creating a national space program based on industry development, technology and innovation. The ASRA concept was, however, declined like its predecessors, the government view being that there was no case for Australia to form a space agency, and that the tracking station relationship between NASA and the Department of Supply was totally adequate.

In 1974, because of its previous involvement with ELDO, Australia was invited to become a full member of the European Space Agency, which was being formed following the collapse of the ELDO program in the early 1970s. It was the only non-European country to which such an invitation has ever been extended. Although the Whitlam Government declined to take up the ESA offer, as it was not interested in furthering Australian civil space activities at that time, it did, however, reserve the right to take up ESA membership at a future date.

It was not until 1983 that Australia actually rescinded its rights to ESA membership, ironically declining to join the Agency just at the time when local space science and industry were undergoing a significant rejuvenation as a result of the Starlab project and the establishment of Aussat. ESA membership could have been very beneficial to Australia in the 1980s, both because of the opportunities it would have presented to Australian

space science through its own active science program, and because of the potential work that Australian space industry might have obtained as a result of ESA's policy of providing project contracts to member countries on the basis of their financial contribution to that project.

Since 1983, ESA has made further invitations to Australia to join the Agency, albeit on the basis of Co-operating State status (the same as enjoyed by Canada). Despite the continued potential for local space industry to receive contracts commensurate with any Australian project funding, both Liberal and Labor governments have declined to take up these invitations, reluctant to provide funding to ESA's mandatory programs (to which all members must contribute and from which there might not be a particular direct benefit to Australia), in addition to specific projects to which Australia might wish to contribute.

Although both the Whitlam and Fraser Governments drew back from supporting local space science and technology projects and were unwilling to make a major financial commitment to international space programs, space applications such as meteorology, remote sensing and communications played an increasing role in Australia's national development in the 1970s. This practical use of space was funded by a number of government departments to meet specific national needs, which could be assisted by space applications. The growing use of space applications would eventually initiate the next phase of Australian space engagement in the 1980s.

During the 1970s, many ground sector applications programs, such as the reception and analysis of Landsat imagery, also found support through general science funding and the CSIRO, while the construction of satellite telecommunications ground stations by OTC was supported due to the national requirement for international telecommunications. This enabled Australian space science and industry to develop strengths in certain space disciplines, which came to the fore in the 1980s when funding for space activities became more readily available.

Space Science and Industry Lead a New Phase of Space Engagement

The development of the Starlab project and the decision to establish the Aussat satellite communications system encouraged a dramatic revival of Australian interest in space in the 1980s. The opportunities that these projects afforded Australia paved the way for a rejuvenation of local space science and industry resulting in the eventual development of Australia's first formalised national space policy and the creation of its first national space instrumentalities (though neither could truly be called a 'space agency'). In the light of government industry and technology policy at the time, Starlab was seen as a way for Australian industry to regain its lost space qualification status and encourage the development of new high-technology industries: Starlab would provide a means of utilising the offset provisions for Aussat to promote the growth of an Australian space industry.

1984 was probably the most important year in this new phase of Australian re-engagement with space. Although, ironically, it saw the cancellation of the Starlab project, it also saw the fruits of the impetus it had provided. The First National Space Symposium, held in that year, had more than 200 local and overseas participants. It was instrumental in providing the momentum that eventually led to the development of two government-funded 'space offices': the CSIRO Office of Space Science and Applications (COSSA) and the Australian Space Office (ASO). Also at this time, the Institution of Engineers, Australia, established its National Panel (and later, Committee) on Space Engineering, to consider and promote professional interest in the discipline.

In late 1984, following the Symposium, the CSIRO, which had long had a wide range of space science and applications interests, moved to consolidate its many areas of space activity, establishing COSSA to co-ordinate all of its space projects. COSSA's annual budget of around $6 million provided for remote sensing systems, communications research, radio astronomy and other related fields, with links to many international space programs. COSSA, as outlined in earlier chapters, was particularly active in space applications projects, especially those related to environmental monitoring. It initiated the development of sensors and instruments for spacecraft, such as the Along Track Scanning Radiometer components that Australia contributed to the European ERS 1 and 2 satellites, and their successor, Envisat.

Above A mosaic of remote sensing images from the ERS 2 satellite, which carried the Along Track Scanning Radiometer instrument to which Australia contributed. (Image courtesy of ESA)

A Space Policy for Australia

Following the First National Space Symposium there were calls for Australia to develop a comprehensive national space policy to replace the ad hoc decision-making that had previously affected Australian civil space activities. The Australian Academy of Technological Sciences was commissioned by the Hawke Government's Minister for Science, Barry Jones, to prepare a report on the potential of space science and technology in Australia.

In 1985, the Academy released its report, *A Space Policy for Australia* (also known as the Madigan Report, after Sir Russel Madigan, the Chair of the Academy's Working Party which developed it). This report considered a number of issues relating to space programs in the nation, including whether Australia had the potential financial, technological and industrial capacities to service a positive space policy. While the report found that the country did indeed have these capabilities, it speculated as to whether the Government had sufficient commitment to support a national space program – an astute observation that was borne out by subsequent events. It also noted that no formal space policy existed, nor any agency capable of focusing national efforts on space affairs (even though the establishment of COSSA had at least focused CSIRO efforts).

In its extensive review of national space activities, which included a comparison with the space activities of other nations (particularly Canada, whose level of space engagement has regularly been held up as an example Australia should seek to emulate), the report concluded that Australia should establish a national civil space policy under Government leadership, emphasising the space-related ground infrastructure sector, particularly involving remote sensing. The study also supported international collaboration on space projects, especially with neighbours in South-East Asia, and further urged that government-funded research and development contracts be placed with Australian industry in order to establish local subcontractor or prime-contractor status for space hardware. It suggested an initial government commitment to spending $100 million over five years towards suitable programs, which would be co-ordinated under an independent national statutory space authority.

The Australian Space Board and Office

In response to the Madigan Report, in 1986 the Hawke government established the Australian Space Board (ASB) to oversee a National Space Program (NSP). Although the Madigan Report had recommended that a statutory authority be established to manage the National Space Program, the ASB was constituted as an advisory body only, located within the then Department of Industry, Technology and Commerce (DITAC). The major goal of the National Space Program was industry focused, aimed at encouraging greater involvement by Australian industry in space research and development activities to promote the development of commercially viable industries based on space technologies. The ASB's role was to act as an advisory body in the formulation of a national space policy. Its main functions were to facilitate the development of Australian space science and technology, identify industry opportunities, and establish international programs of collaboration.

To support the ASB, the Australian Space Office (ASO) was established in 1987 to act as its secretariat and conduct the daily activities necessary for the co-ordination and management of the National Space Program. Situated within DITAC, the ASO's role was to co-ordinate the development of commercially viable industries based on space technologies, which were export-oriented and internationally competitive. The ASO would develop five sections: the Space Projects section, which oversaw the contracts and specifications for NSP-funded space projects; the Launch Services Section, responsible for monitoring the progress of the many private sector launch facility proposals that arose during the late 1980s and 1990s; the Space Policy Section, responsible for the development of Australian space policy and the programs necessary to achieve policy objectives; the NASA Administration Section, which handled the administrative, contractual and financial arrangements associated with NASA's activities in Australia; and the Canberra Deep Space Communications Complex (the formal title of the Tidbinbilla space tracking station), which was important enough to warrant a separate ASO section.

Although the ASB worked closely with COSSA, the National Space Program it managed actually represented only a small part of Australia's overall space-related activities at this time: Aussat/Optus, OTC/Telstra, the Bureau of Meteorology and other government departments, Defence space activity, communications and many other space applications activities continued to operate outside the purview of the NSP and did not come under its jurisdiction. This fragmentation of responsibility, coupled with a lack of overall co-ordination in the management of Australia's space activities, was strongly criticised for creating duplication and overlap of activities and responsibilities — a situation, and a criticism, that persists until this day. It was, and remains, also frustrating for potential international collaborators, due to the difficulty of determining who was responsible for particular space activities and who was responsible for funding particular types of projects (since funding for many Australian space activities also lay outside the ASO's funding responsibilities).

As its placement within DITAC made clear, the initial objectives of the ASO were focused very much on the development of Australian industry and the fulfilment of Australia's space responsibilities under international space agreements. The ASO directed its small amount of available funding toward space industry development projects and a range of studies examining potential areas of opportunity for niche-market development of Australian space-related products. It sought to promote industry through its Space Industry Development Strategy and the establishment of Space Industry Development Centres, linking universities and the commercial sector in the development of promising technological innovations.

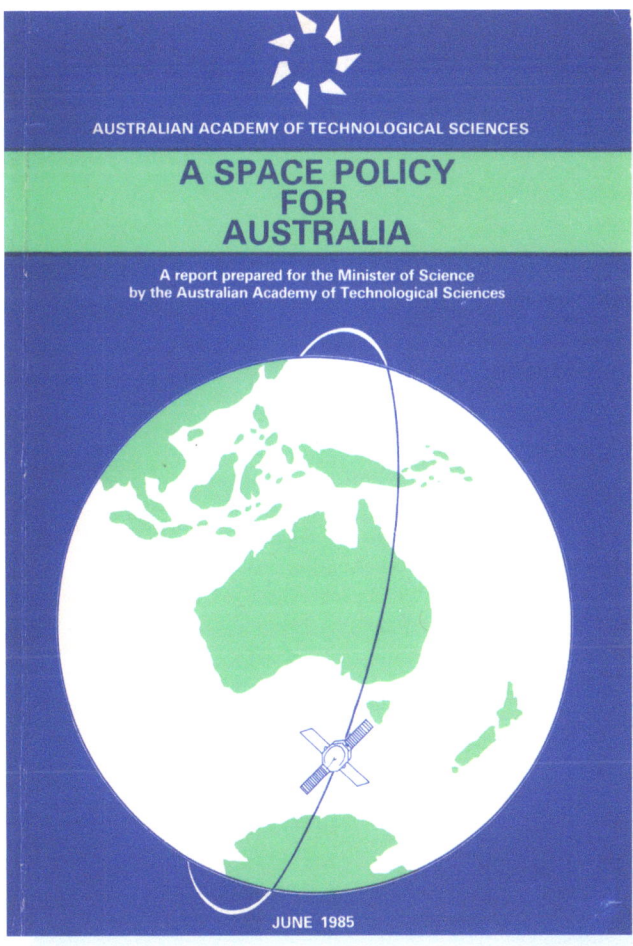

RIGHT The Madigan Report put forward the basis for a comprehensive national space policy. Although not fully adopted by the Government, it did lead to the formation of the first National Space Program. (Courtesy of the Australian Academy of Technological Science and Engineering)

Within the National Space Program, science objectives and 'national good' interests in space products and services were perceived to have a lower priority than industry, which provoked criticism from the Australian space science community that the ASO did not devote sufficient resources to space science programs, despite the fact that space science always had a greater proportion of the space budget than the 10 per cent minimum recommended by the 1992 Curtis Report. Space science programs (such as the Lyman/Endeavour and ATSR projects) were initially part of the NSP, being transferred from the Department of Science under which they had commenced to the ASB on its creation. While many of the projects that the ASO supported were science driven, each was tailored to have relevance to the development of Australian space industry.

To address the concerns of the space science community, the ASO funded the Australian Academy of Science's production of the 1988 *Ready for Launch: Space Science in Australia* (Cole) report, which recommended consistent government support for an active program of space science. It outlined a five-year plan, with special funding for programs in ultraviolet astronomy, airborne science, and remotely sensed data use, all areas in which Australia had expertise. The report also urged specific funding for the establishment of a space and astronomy data storage facility, and the design, construction and launch of an Australian science and applications spacecraft, as well as support for space science and education programs.

The ASO sought to redress the perceived industry bias of the National Space Program by promulgating the Balanced National Space Program in 1991, which refocused policy directions to include opportunities for public goods and services and national benefits from the use of space, while at the same time continuing to address the requirements of industry. However, its lack of a direct response to the Cole Report drew criticism from some sections of the astronomy community, for failing to promote the development of space astronomy and the creation of an associated database. By 1992, these astronomers were calling for the creation of a completely independent and professional space agency that would directly employ engineers and scientists, oversee contracts and maintain a space database with international collaboration. This call would be repeated in the Academy of Science's submission to the 1995 Interdepartmental Committee on International Space review.

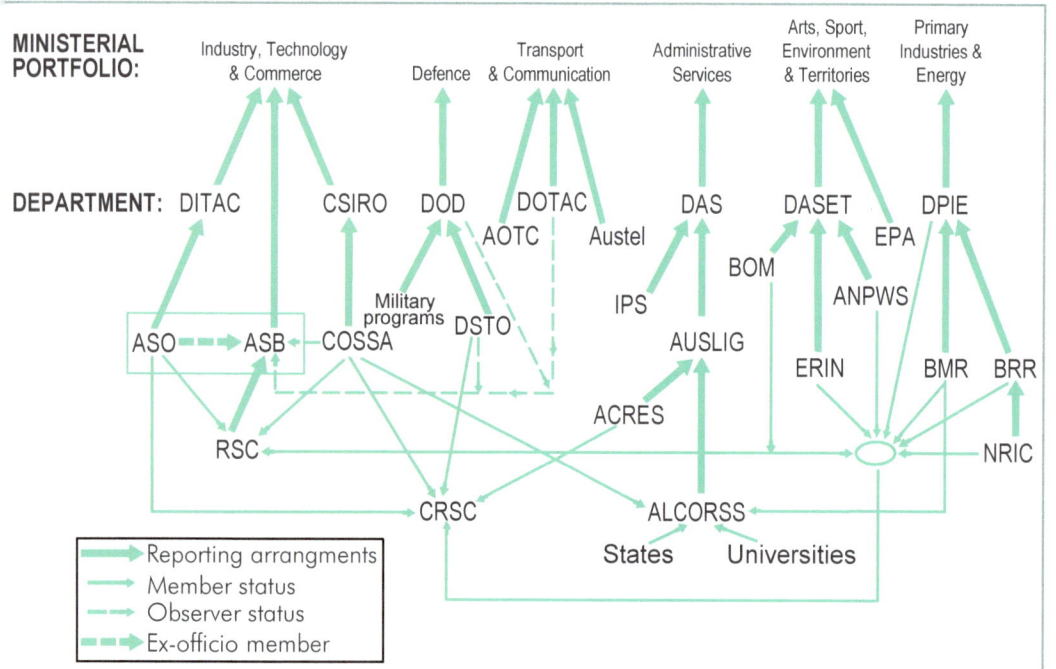

Above The complex organisational arrangements existing in 1992 between different government departments controlling space activities are outlined in this chart, derived from the Curtis Report.

ALCORSS: *Australian Liaison Committee on Remote Sensing by Satellite (observers: ASB, ASO)*
ANPSW: *Australian National Parks & Wildlife Service*
Austel: *Australian Telecommunications Authority*
BMR: *Bureau of Mineral Resources, Geology & Geophysics*
BOM: *Bureau of Meteorology*
BRR: *Bureau of Rural Resources*
CRCS: *Commonwealth Remote Sensing Committee*
ERIN: *Environmental Resources Information Network*
IPS: *Ionospheric Prediction Service*
NRIC: *National Resources Information Centre*
RSC: *Remote Sensing Committee*

Although the Madigan Report recommended that the National Space Program receive funding of approximately $20 million per year over a five-year period in order to develop an effective level of space activity, the ASB only received successive budgets of around $3-$5 million per year, extremely low by international standards, and allowing little more than coverage of ongoing commitments: it was wholly inadequate to support a major space project, such as the revitalisation of Woomera as a LEO launch facility, which, as outlined in Chapter 8, was strongly supported by the Australian space industry at the time.

The percentage of Gross National Product (GNP) actually devoted to Australia's formal space program at that time was significantly smaller than that allocated to space activities in other countries, even those nations considered far less developed than Australia: this was perceived by critics of Government space policy as indicative of a lack of Government commitment to space involvement. However, at the same time, indirect expenditure on space-related systems by various government agencies amounted to around $100 million each year, mostly spent overseas on purchases such as the procurement and launch of satellites for the Aussat/Optus system, meteorological and communications satellite ground receiving stations, in addition to the ongoing operation of remote sensing satellite ground stations and processing facilities not under ASB control.

THE 1990s: REVIEW AND REASSESSMENT

International Space Year, 1992, saw the release of three government reports on space activities and opportunities in Australia that were to have an important effect upon the country's space policy and the National Space Program. The first of these reports, the Senate Standing Committee on Transport, Communications and Infrastructure's Developing Satellite Launching Facilities in Australia and the Role of Government, considered the feasibility of the Cape York Space Port and Woomera revival proposals and made positive recommendations for non-financial government support to both these projects, should they proceed and prove to be financially viable. These recommendations would lie behind future government reluctance to directly support spaceport proposals and their exclusion from Australia's Satellite Utilisation Policy in 2013.

More significantly, the five-year review of the National Space Program, as suggested in the Madigan Report, was undertaken by two independent groups, both assessing the performance of the NSP from very different perspectives. The first review, commissioned by DITAC to assess the performance of the NSP and make recommendations for its future direction, was undertaken by a three person Expert Panel, consisting of recognised space or technological experts (one foreign and two Australian), with a wealth of experience in both government and private sector space activities. Concurrent with that study was an economic evaluation of the NSP conducted by the Bureau of Industry Economics (BIE).

The Expert Panel report, *An Integrated National Space Program* (also known as the Curtis Report), found that, despite criticisms in regard to co-ordination of activities among different agencies, lack of support for space science and small budgets, the NSP and the ASB/ASO had performed well, particularly in developing Australian space industry capability significantly within a relatively short period of time. Noting the criticisms in regard to co-ordination of activities, the report recommended that a high-level Australian Space Council be established to formulate space policy and set priorities for a revitalised National Space Program. It also advocated a threefold increase in the space budget over three years, emphasising commitment to remote sensing for environmental monitoring, mobile communications systems, aspects of launch services, space science, and continued growth of the space industry.

A particular weakness of the NSP noted by the Curtis Report was that defence-related space activities were outside its purview, resulting in the loss of the benefits and efficiencies that could be derived from an integrated space program without overlapping projects in the civil and defence sectors.

Evaluating the NSP from an economic perspective, however, the BIE Report, *An Evaluation of the National Space Program*, found that there was no economic case for an independent National Space Program. Although the BIE Report concurred with the Curtis Report on the importance of remote sensing and strong funding for a national remote sensing program, it recommended that future funding for space industry development be provided through general industry development support programs, while space science should become a possible priority for funding within general science programs. As such, this report's thrust was at odds with the philosophy expressed by the Curtis Report, and was criticised by the space community for not including on its committee anyone with specific space experience or expertise.

Faced with two quite different views, the Keating Government was required to make a decision as to which of the reports' recommendations it would follow in establishing the direction of Australia's future space policy, and this period of uncertainty led to the establishment of the Australian Space Industry Chamber of Commerce (now SIAA), to promote space industry interests. The government ultimately accepted the Curtis Report's recommendations, insofar as it replaced the Australian Space Board with a statutory body, the Australian Space Council (ASC), supported by an enhanced Space Office within DITAC. The Council drew its members from industry, the research community and government departments with a strong space interest, reflecting a wider membership than the previous ASB. However, despite the establishment of the ASC, the budget recommended by the Curtis Report was not forthcoming, and the new Integrated National Space Program continued to operate with little more funding than its predecessor.

Although the National Space Program was not due for its next review until around 1997, in 1995 the Government announced a new evaluation of the NSP by an Interdepartmental Committee on International Space, composed of senior bureaucrats from those government departments with involvement in space activities. A submission to this review by the Institution of Engineers, Australia, suggested that a local space industry had failed to thrive because Australian companies lacked significant domestic space projects on which to build a critical mass of expertise and support.

By this time, Australia was paying over $600 million annually for overseas space services (in the estimation of the Institution of Engineers), although it also enjoyed essentially free access to the data provided by the meteorological, remote sensing and scientific satellites launched by other nations. As previously mentioned, a constantly expressed concern was that this free access might be lost unless Australia 'bought in' to these programs by providing funding and/or instruments, or becoming an industry sub-contractor, to future international applications of satellite programs.

While the Interdepartmental Committee reported favourably on the economic basis of national space capability, the National Space Program, the ASC and the ASO were all terminated in June 1996 by the new Howard government. They were replaced in August 1996, by new administrative arrangements for space activities, coupled with the announcement of a new small satellite program, to be known as FedSat (Federation Satellite). This program was designed to build on existing national research experience and industry capabilities through the production of a small demonstrator satellite that would, at the same time, celebrate the centenary of Australia's Federation.

Under the new government plans, the ASO was replaced by a small Space Policy Unit within the then-Department of Industry, Science and Tourism (DIST), while COSSA was restructured to undertake the FedSat project, for which there was a funding pledge of $20 million. FedSat would be developed by a new Cooperative Research Centre for Satellite Systems (CRCSS), which would be created by transferring the core staff of COSSA to the new centre: the remaining section of COSSA, focused on remote sensing, was attached to the CSIRO's Earth Observation Centre. Although the CRCSS, which commenced operation on 1 January 1998, was intended to have a broader role than FedSat, covering the long-term strategic operational and commercial role for satellites, and, as outlined in Chapter 8, achieved modest success in this aim, it was closed at the end of 2005 when its funding was not renewed.

In early 1998, to provide a legislative framework under which the various proposals for commercial launch facilities in Australia (discussed in Chapter 8) could be regulated, the government passed the Space Activities Act, the world's first formal legislation specifically covering commercial space launch operations. It was also the first domestic legislation in the world to define, for its purposes,

the point at which 'space' can be considered to begin, this being taken as 100km above the Earth's surface.

The Act was designed to ensure that all launch activities for which the Australian Government could be considered responsible under international law took place in accordance with that law. It created regulations governing matters of licensing, insurance, launch site approval, space object registration, range safety, fees and liability for commercial space activities carried out in Australia or by Australian nationals or entities overseas. The Act specifically required launch operators to indemnify the Government against its international liability for loss and damage and established stringent launch safety standards and a detailed launch insurance cover regime. To oversee the requirements of the Act, the Space Licensing and Safety Office (SLASO) was established within DIST, responsible for the issue of launch licences for activities carried out both in Australia and overseas.

Lost in Space? Australia and Space at the Dawn of the Twenty First Century

Despite the promulgation of the Space Act, and it support for FedSat as a demonstration of the capabilities of Australian space industry, the Howard government considered the space sector to be similar to other high-technology industries and did not see the necessity of providing specific support to the development of a local space industry. Consequently, as the Twenty First Century began, the lack of a co-ordinating body or a national space program meant that many companies that had developed space capabilities during the space industry development push of the 1980s and 90s either folded or turned to other industries in order to survive commercially. The commercial LEO launch proposals in the 1990s had also been abandoned, due to the demise of the huge mobile communications satellite constellations on which their economic viability was predicated (which were rendered un-necessary by the growth of the terrestrial phone-cell network), while the none of the equatorial spaceport proposals had come to fruition, due to their being economically marginal at best.

The then-Department of Industry, Tourism and Resources (DITR) retained responsibility for civil space, acting as Chair of the Australian Government Space Forum, which also included the Departments of Defence, Communications, Science and Education. The function of this forum was to co-ordinate with other government agencies involved in space, exchange information about policies, programs and activities and encourage a collaborative approach to space issues and programs.

Above Commercial spaceport projects like the Asia Pacific Space Centre, seen in this artist's conception, prompted the creation of the Australian Space Act to regulate launch activities in Australia. (Illustration created by Asia Pacific Space Centre)

Following a letter by Australian astronaut Andy Thomas to Prime Minister Howard, accompanied by efforts from NASA to engage Australia in the International Space Station program, an International Space Advisory Group (ISAG), chaired by Paul Scully-Power, the only other Australian-born person to have flown in space, was appointed in mid-2001 to identify opportunities for Australian involvement in the International Space Station (ISS) and other international space programs. The ISAG report recommended support for an Australian space industry, but it was met with indifference by Government and no action was taken.

In 2003, DITR developed a Policy Framework for Space Engagement, which focused on ensuring access to space services, supporting science and space-related research, promoting the growth of private sector industries and safeguarding national security. The inclusion of national security in the policy framework was a recognition of the importance of space-based assets in this area and a direct response to the tense geopolitical situation of the time: it marked the first time that defence and security-related matters had been formally incorporated into a national space policy.

Despite the policy framework's support for space science and industry development, it continued the existing Government approach, specifically ruling out the establishment of a centrally funded space agency or national space program and declining to provide dedicated support for space industry, on the basis that space industry development issues were similar to those faced by other high technology areas and were, therefore, adequately addressed by existing general industry development programs.

The Australian Government Space Engagement Policy Framework was issued in late 2003, being subsequently updated in 2004, 2006 and by the Rudd Government in 2008. This document signalled a more pragmatic approach by Government to Australian space activities, where future engagement with space would be driven by market forces, particularly those areas in which Australia had a competitive advantage. It was anticipated that much of Australia's space activity would derive from the provision of national technological expertise and ground station tracking services.

The Space Engagement Policy Framework drew a response from South Australian Liberal Senator Grant Chapman, Chairman of the Government's Industry and Resources Committee and a long-time proponent of a national space policy, who believed that the lack of a co-ordinated national space policy and space program was not in Australia's best interests. In 2005, drawing together a Space Policy Advisory Group composed of academic, industry and other space experts, Chapman prepared a submission to the Prime Minister and Cabinet, *Space: a Priority for Australia,* which was presented to the Howard Government in January 2006.

This document recommended that the government acknowledge the strategic national importance of space to Australia and formulate a national space policy that would ensure Australia's continued access to critical space-based services and data; increase Australia's space capabilities; and ensure the nation a voice in the international dialogue on space issues. It suggested that responsibility for co-ordinating the Australian strategic space policy framework be assigned to an agency with experience in managing broader national issues, such as the Department of Prime Minister and Cabinet.

Chapman's support for a national space policy provoked considerable debate within parliamentary and space community circles, both in the civil and Defence sector and resulted, in 2008, in the Senate requiring its Standing Committee on Economics to produce a report on the current state of Australia's space science and industry sector. The outcome of this review was the report *Lost in Space? Setting a New Direction for Australia's Space Science and Industry Sector,* which, recognising that Australia was the only major nation in the Organisation for Economic Co-operation and Development (OECD) not to have a national space agency, put forward six recommendations for the gradual establishment of a national space policy and a national space agency:

1. *That Government allocate resources to the existing Space Policy Unit to enable the establishment of an Australian government Space Information Portal, a website which would provide information on government programs and contacts, and links to Australian companies working in the space industry.*

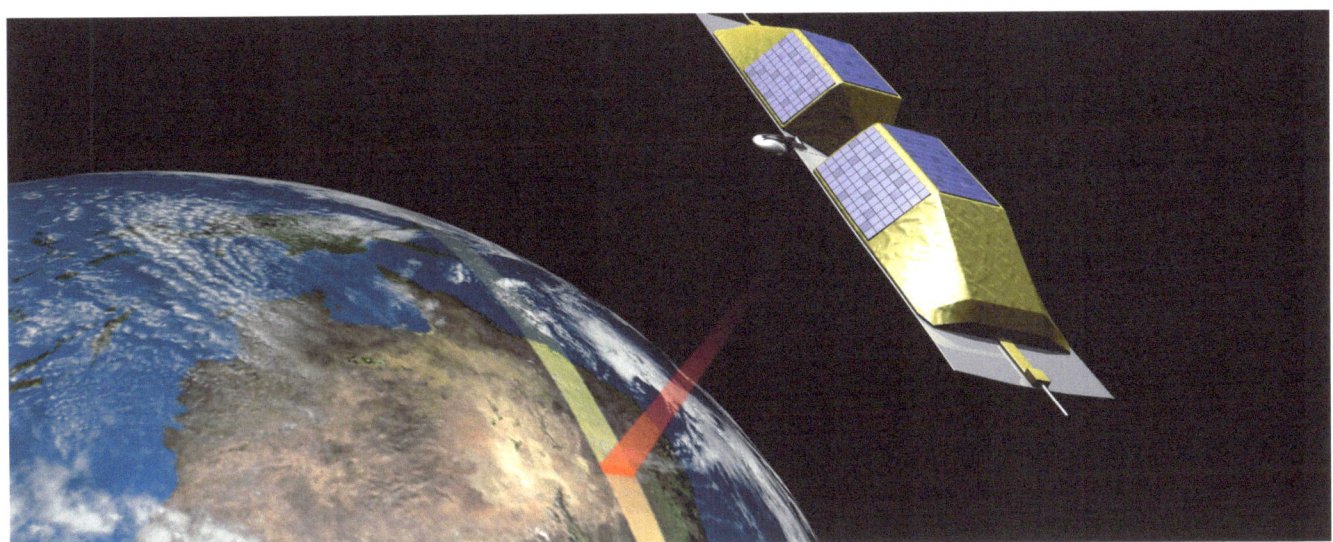

ABOVE *One of the ten projects awarded funds in the ASRP's 'space science and innovation' stream was the Garada project at the Australian Centre for Space Engineering Research. Its focus was developing a space-based system for measuring the moisture in Australian soils. (Illustration courtesy of Australian Centre for Space Engineering Research, University of NSW)*

2. *That immediate steps be taken to co-ordinate Australian space activities and reduce over-reliance on other countries in the area of space technology.*

3. *That the Government establish a unit to co-ordinate Australian space activities, including those in the private sector, with a proper balance between industry and government involvement.*

4. *That the Government initially establish a Space Industry Advisory Council comprising industry representatives, government agencies, defence, and academics and chaired by the Minister for Innovation Industry Science and Research or his representative.*

5. *That the Advisory Council, as a precursor to the establishment of the space agency, undertake audits, assessment and analysis of: Australia's space activities; the strengths, weakness, opportunities and threats in relation to an emerging Australian space industry; the initial, medium and long-term priorities for an Australian space agency; the benefits to Australia of improved international collaboration, including membership of international space bodies; critical performance areas such as research and technological development. The Council should also develop a draft strategic plan for the establishment of a space agency and the most appropriate form of that agency.*

6. *That any Australian Space Agency re-assess the case for Australia becoming more closely linked to an international space agency.*

The committee's unanimous report ultimately concluded:

> 'The committee believes it is not good enough for Australia to be lost in space. It is time to set some clear directions. The Australian government should have a space policy and, like most other comparable countries, an agency to implement it. The global space industry generates global revenues of around US$250 billion per annum, and Australia should be playing a larger role.'

THE AUSTRALIAN SPACE RESEARCH PROGRAM

Despite the then-Department of Industry, Innovation, Science, Research and Tertiary Education (DIISRTE) estimating that Australian space-related activities comprised some 631 organisations with total revenue of between $1-2.2 billion per annum, the Rudd Labor Government did not proceed with the *Lost in Space?* Report's clear call for the establishment of a national space agency. It did, however, respond to the Report's recommendations about the need for the Australian government to have a greater role in co-ordinating and leading the development of space policy and the space sector by re-invigorating the Space Policy Unit in DIISRTE and establishing the Australian Space Research Program (ASRP) in 2009. As outlined in earlier chapters, the objective of this program was to develop Australia's niche space capabilities, through support for space-related research, innovation and skills in areas of national significance or excellence.

The ASRP provided $40 million in three-year project grants in two 'streams': space science and innovation projects, to support collaborative projects for niche space capability development; and space education development projects to support student, or student-focused, projects and space education activities, including international education opportunities and the establishment of national space education programs and centres of expertise for space education. The ASRP funding was allocated to 14 projects over four rounds of funding (four in education and ten in space science and innovation), achieving direct co-investment from industry of $39.1 million and establishing a number of ongoing projects.

An independent assessment of the ASRP in 2015 concluded that the program had represented value for money, had been delivered efficiently, and achieved its objectives whilst continuing to realise benefits to the space sector. However, despite providing strong evidence that government funding can play a critical role in space industry development, the ASRP was terminated under the Abbott Government in 2014.

Australia's Satellite Utilisation Policy

In developing the first formal government space policy since the 1980s, the Space Policy Unit formulated a set of Principles for a National Space Industry Policy in 2011, which ultimately became Australia's Satellite Utilisation Policy, approved by the Gillard Government in 2013. This document outlined seven principles upon which Australia's space industry policy would be based: a focus on space applications of national significance; assured access to space capability; increased and strengthened international co-operation; support for a stable space environment for satellite operations; improved domestic co-ordination; support for innovation, science and skills development; and the enhancement and protection of national security and economic wellbeing.

The policy's stated goal was to achieve on-going, cost-effective access to critical space capabilities while delivering the benefits of improved national productivity, better environmental management, national security and safety (through improved disaster management), a smarter workforce and equity of access to information and services. At the same time, it specifically ruled out Australian participation in launch services, human spaceflight and planetary exploration, and did not support the establishment of any type of space agency. Instead, it called for the establishment of a Space Coordination Committee, as a co-ordination and priority-setting mechanism for civil space activities.

The policy was criticised by many in the Australian space community for its narrow focus on satellite utilisation, while failing to address other aspects

Below The complex Australian Government co-ordination framework for civil space activities, as presented in the 2013 Australia's Satellite Utilisation Policy. (Image courtesy of Department of Industry, Innovation and Science)

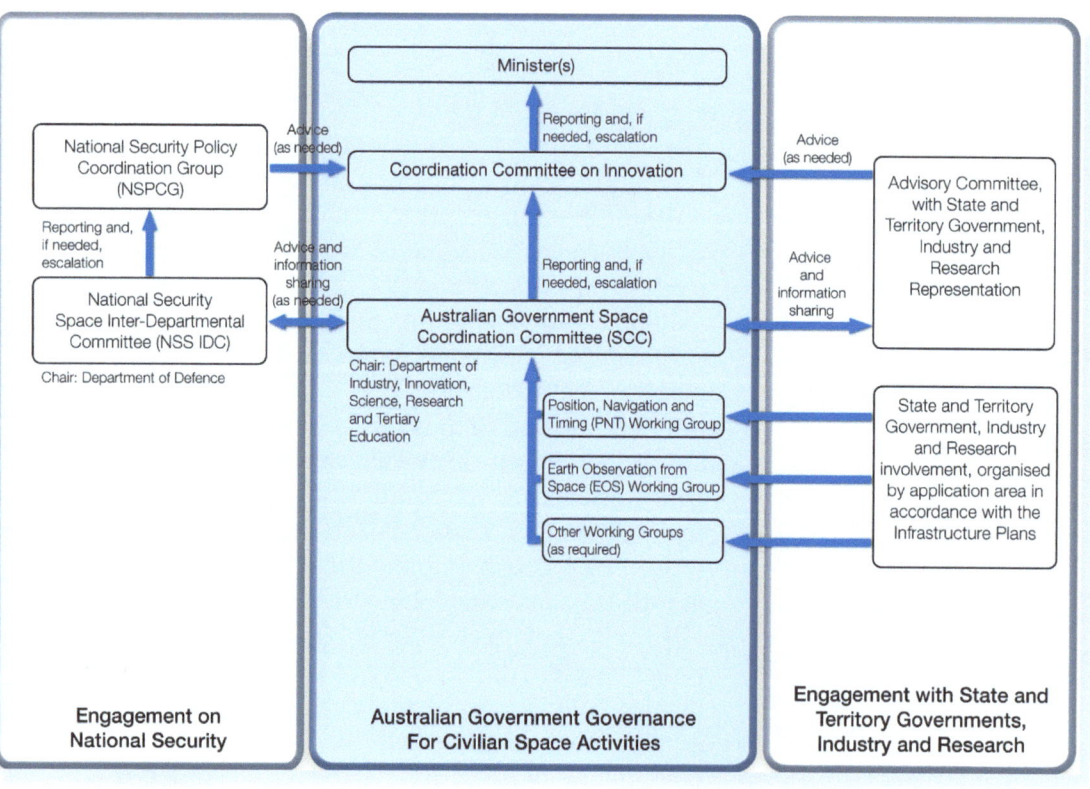

of Australian space activities. It did not offer: a mechanism for national oversight and program management through the creation of a space agency or other co-ordinating body; any clear and measurable strategy for growing the space industry sector or any strategy or follow through to promote the commercialisation of technology and knowledge gained from the Australian Space Research Program. It was also criticised for not addressing the needs of the space science community, which had put forward its own *Decadal Plan for Space Science* in 2010, and for not presenting a national strategy for capability development, including promoting STEM studies at school level and education, training and professional development at the tertiary level.

AT THE SIXTIETH ANNIVERSARY OF THE SPACE AGE: A NEW BEGINNING?

In 2017, sixty years after the dawn of the Space Age, Australian space activity once again faces a period of uncertainty, with the status of the *Satellite Utilisation Policy* unclear, following the change of government in September 2013. At the time of writing, Australia remains the largest OECD nation without a space agency and the only G20 nation to lack one. Where once Australia was considered among the active spacefaring nations—with the launch of WRESAT fifty years ago—the country is now perceived internationally as 'punching well below its weight' in space, given the size of its economy. Despite being a sophisticated user of space-based services, Australia is generally considered a passive 'consumer' of space products, rather than an active participant in global space activity.

The Australian space sector produces annual revenues of around $3-$4 billion (only 8 percent of which is generated from exports) and has an estimated workforce of between 9,500 and 11,500 full time equivalents. These figures represent approximately 0.8 percent of the global space economy, which was estimated to be worth US$330 billion in 2014, and indicate that the decentralised approach to space activities taken by successive governments over the past six decades (despite repeated reports recommending the establishment of some form of national co-ordinating body for Australian space activities) and the low levels of direct government investment in the development of space science and industry capabilities, have not been the most effective or successful strategy for the nation to follow.

The Government is currently reviewing the Australian Space Activities Act, whose complex regulatory requirements have proved onerous to universities, agencies and small entrepreneurial start-ups seeking to take advantage of the much cheaper access to space offered by the recent development of cubesats, which usually launch as subsidiary payloads with other, larger satellites. A simplification of the Space Act's requirements for smallsat users would enable access to space for a broader section of the community and offer new opportunities for exciting, hands-on STEM education projects.

There are once again calls from the civil space community for a comprehensive national space program, overseen by a co-ordinating agency, with an adequately funded budget that will effectively promote Australian space science and industry. The Space Industry Association of Australia has submitted a White Paper, *Advancing Australia in Space*, to Government outlining a permanent and sustainable national civil space program that addresses Australian space priorities, industry growth, civil space regulation and international legal responsibilities, capability development (from STEM to the space workforce), and the steps to establish an Australian Space Agency, reporting direct to the responsible Minister.

As previous chapters in this book have shown, Australia has had a proud history of space involvement, and today possesses the scientific, engineering, technological and industrial skills to assume a much larger role in global space activities, with concomitant economic, social, scientific, educational, industrial and national security benefits. Will Australia continue to remain 'lost in space', or will the government respond to the groundswell of support for a national space program, overseen by an Australian Space Agency, and launch this country into a new and vital phase of space engagement, enabling Australia to permanently become a spacefaring nation?

List of Acronyms

AATSR	Advanced Along Track Scanning Radiometer
ABLS	Australian Balloon Launch Station
ABM	Anti-Ballistic Missile system
ACA	Australian Centre for Astrobiology
ACRES	Australian Centre for Remote Sensing
ACSER	Australian Centre for Space Engineering Research
ACT	Australian Capital Territory
ADEOS	Advanced Earth Observation Satellite
ADF	Australian Defence Force
ADFA	Australian Defence Force Academy
ADS	Australian Department of Supply
ADSCS	Australian Defence Satellite Communication Station
ADSS	Australian Defence Scientific Service
ADT	Australian Department of Trade
AETHERS	Advanced Equipment to Handle ERS
AFRL	US Air Force Research Laboratory
AGO	Australian Geospatial-Intelligence Organisation (originally DIGO)
AGWAC	Australian Guided Weapons Analogue Computer
AITC	Advanced Instrumentation Technology Centre
ALSEP	Apollo Lunar Surface Experiment Package
ALFLEX	Automatic Landing Flight Experiment
ALV	Australian Launch Vehicle
AMSAT	Radio Amateur Satellite Corporation
ANCOSPAR	Australian National Committee on Space Research
ANU	Australian National University
AOTC	Australian and Overseas Telecommunications Corporation, formed by the merger of the Overseas Telecommunications Corporation (OTC) and Telecom Australia in 1992- now Telstra
APS	Atmospheric Pressure Sensor
APSC	Asia Pacific Space Centre
ARC	Aggregation of Red Cells experiment
ARDU	Aircraft Research and Development Unit (RAAF)
ARPA	Advanced Research Projects Agency
ARS	Australian Rocket Society
ARTVS	Australian Rocket Test Vehicle Simulator (pronounced 'Artus')
ASB	Australian Space Board (replaced in 1993 by the ASC)
ASC	Australian Space Council
ASD	Australian Signals Directorate (formerly DSD)
ASERA	Australian Space Engineering Research Association
ASG	Australian Spaceport Group
ASICC	Australian Space Industry Chamber of Commerce
ASO	Australian Space Office
ARIES	Australian Resource Information and Environment Satellite
ASRA	Australian Space Research Agency
ASRI	Australian Space Research Institute
ASRP	Australian Space Research Program
ATNF	Australia Telescope National Facility
ATS	Applications Technology Satellite
ATSR	Along Track Scanning Radiometer
AUSLIG	Australian Surveying and Land Information Group
AUSROC	Australian Rocket (ASRI sounding rocket)
BAeA	British Aerospace Australia
BIE	Bureau of Industry Economics
BIOS	Biological Satellite (mission name)
BIOSAT	Biological Satellite (generic term)
BOM	Bureau of Meteorology
BRIAN	Barrier Reef Image Analysis (BRIAN)
CAAG	Cotton Aero-dynamic Anti-G suit
Capcom	Capsule Communicator
CASS	CSIRO Astronomy and Space Science
CDSCC	Canberra Deep Space Communications Complex
CECLES	Conseil européen pour la construction de lanceurs d'engins spatiaux (French name for ELDO)
CIA	Central Intelligence Agency
CNES	Centre National d'Etudes Spatiales (French National Centre for Space Studies)
COMSAT	Communications Satellite. Also, Communications Satellite Corporation
COPUOS	Committee on the Peaceful Uses of Outer Space
COSPAR	Committee on Space Research

COSPAS	Acronym derived from the Russian Cosmicheski Spasetel (Space Rescuer)	DSTO	Defence Science and Technology Organisation, formerly Defence Research Centre (DRC) and Weapons Research Establishment (WRE)
COSSA	CSIRO Office of Space Science and Applications		
CRCSS	Cooperative Research Centre for Satellite Systems	EAST	Eastern Australian Standard Time
		EC0	Educational Cubesat Zero
CSIDA	CSIRO System for Interactive Data Analysis	ELDO	European Launcher Development Organization
CSIRO	Commonwealth Scientific and Industrial Research Organization	Elint	Electronic Intelligence
		ELT	Emergency Locator Transmitter
		EOC	Earth Observation Centre
CYISL	Cape York International Spacelaunch Ltd	EOS	Electro Optic Systems
CYSA	Cape York Space Agency	Envisat	Environmental Satellite
DCA	Department of Civil Aviation	EPIRB	Emergency Position Indicating Radio Beacon
DIGO	Defence Imagery and Geospatial Organisation (now AGO)		
		ERS	European Remote Sensing Satellite
DIISRTE	Department of Industry, Innovation, Science, Research and Tertiary Education	ERTS	Earth Resources Technology Satellite
		ESA	European Space Agency
DISIMP	Device Independent System for IMage Processing	ESOC	European Space Operations Centre
		ESRO	European Space Research Organization
DIST	Department of Industry, Science and Tourism	Estrack	ESA Tracking Network
		ET	Extraterrestrial
DITAC	Department of Industry, Technology and Commerce	Eumetsat	European Meteorological Satellite
		EVA	Extravehicular Activity
DITR	Department of Industry, Tourism and Resources	FDP	Fast Delivery Processor
		FedSat	Federation Satellite
DLR	Deutsches Zentrum für Luft- und Raumfahrt e.V. (German Aerospace Centre)	FUSE	Far Ultraviolet Spectroscopy Explorer
		GAMS	Global Atmospheric Methane Sensor
		GASCAN	'Get Away Special' Canister
DOR	Differential One-way Ranging	GCF	Ground Communications Facility
DORIS	Doppler Orbitography and Radiopositioning Integrated by Satellite (in French, Détermination d'Orbite et Radiopositionnement Intégré par Satellite)	GEO	Geostationary Orbit
		GEO	Group on Earth Observation
		GEOS	Geostationary Earth Observation Satellite
		GMS	Geostationary Meteorological Satellite
DRCS	Defence Research Centre Salisbury, formerly the Weapons Research Establishment (WRE), later the Defence Science and Technology Organisation (DSTO)	GPS	Global Positioning System (see also, NAVSTAR-GPS)
		HACBSS	Homestead and Community Broadcast Satellite Service
		HAD	High Altitude Density
DSA	Deep Space Antenna	HARP	High Altitude Research Project
DSCS	Defence Satellite Communications System	HASP	High Altitude Sounding Projectile
		HAT	High Altitude Temperature
DSD	Defence Signals Directorate (now ASD)	HDLT	Helicon Double Layer Thruster
DSIF	Deep Space Instrumentation Facility (original name for DSN stations)	HIBAL	High Altitude Balloons
		HIFiRE	Hypersonic International Flight Research Experimentation
DSN	Deep Space Network		
DSP	US Defence Support Program		
DSS	Deep Space Station	HF	High Frequency (radio)
DST Group	Defence Science and Technology Group (formerly DSTO)	HOPE	H 2 Orbiting Plane (proposed Japanese spaceplane)

List of Acronyms (Continued)

HOTOL	Horizontal Take-Off and Landing
HRV	Hypersonic Research Vehicle
HSFD	High-Speed Flight Demonstration
HyCAUSE	Hypersonic Collaborative Australian/United States Experiment
IBM	International Business Machines
ICBM	Intercontinental Ballistic Missile
IGY	International Geophysical Year
IMAGED	Image-based Analysis of Geographic Data
IMOS	Integrated Marine Observing System
IMP	Interplanetary Monitoring Platform
INMARSAT	International Maritime Satellite Organisation
INSPIRE	Integrated Spectrograph Imaging and Radiation Explorer
INTELSAT	International Telecommunications Satellite Organisation
IMSO	International Mobile Satellite Organization
IRC	International Resource Corporation
ISS	International Space Station
ISU	International Space University
JAXA	Japan Aerospace Exploration Agency
JPL	Jet Propulsion Laboratory
JORN	Jindalee Operational Radar Network
LA	Launch Area
LBA	Long Baseline Array
LEO	Low Earth Orbit
LEOP	Launch and Early Orbit Phase
LFPCD	Large Format Photon Counting Detector
LLR	Lunar Laser Ranging
LRWE	Long-Range Weapons Establishment
LUT	Local User Terminal
Minitrack	Minimum Weight Tracking
MILA	Merritt Island Launch Area
MIT	Massachusetts Institute of Technology
MOBLAS	Mobile Laser Station
MSFN	Manned Space Flight Network
MSG	Mathematical Services Group (a section of the LWRE/WRE)
MTSAT	Multi-functional Transport Satellites
MUAS	Melbourne University Astronautical Society
MUOS	Mobile User Objective System
NASA	National Aeronautics and Space Administration
NASCOM	NASA Communications
NASP	National Aero Space Plane
NAVSTAR	Navigation System using Timing and Ranging-Global Positioning System
NBN	National Broadband Network
NEOG	National Earth Observation Group
NEXST	National Experimental Supersonic Transport
NICTA	National ICT Australia Ltd
NOAA	National Oceanic and Atmospheric Administration
NRL	Naval Research Laboratory
NRO	National Reconnaissance Office (US)
NSA	National Security Agency (US)
NSP	National Space Program
NSSA	National Space Society of Australia
NSW	New South Wales
OCS	Ocean Colour Scanner
OSCAR	Orbiting Satellite Carrying Amateur Radio
OSETI	Optical Search for Extraterrestrial Intelligence
OTC	Originally the Overseas Telecommunications Commission (1946), later Corporation (1989). Merged with Telecom in 1992 to form Telstra.
PACT	Pacific Area Cooperative Telecommunications (network)
PLB	Personal Locator Beacon
PMG	Postmaster General's Department (1975 split into Telecom Australia and Australia Post)
RAAF	Royal Australian Air Force
RAE	Royal Aircraft Establishment (later Royal Aeronautical Establishment)
RASS	Remote Area Satellite Service
READS	Re-entry Air Data System
RF	Radio Frequency
Rockoon	Rocket Balloon
SA	South Australia

List of Acronyms

SAO	Smithsonian Astrophysical Observatory
SARSAT	Search and Rescue Satellite
SatCen	Satellite Centre
SATCOM	Satellite Communications
SCRAMJET	Supersonic Combustion Ramjet
SERC	Space Environment Research Centre
SETI	Search for Extraterrestrial Intelligence
SIAA	Space Industry Association of Australia
SIDC	Space Industry Development Centre
Sigint	Signals Intelligence
SIR	Spaceborne Imaging Radar
SHSSP	Southern Hemisphere Space Studies Program
SLASO	Space Licensing and Safety Office
SLIP	Software for LANDSAT Image Processing
SLR	Satellite Laser Ranging
SLV	Southern Launch Vehicle
SPAN	Solar Particle Alert Network
SPARTA	Special Anti-missile Research Tests, Australia
SPARTAN	Scramjet Powered Accelerator for Reusable Technology AdvaNcement
SPOT	Satellite Probatoire d'Observation de la Terre (Satellite Probe for the Observation of the Earth)
SPU	Space Policy Unit
SSA	Space Situational Awareness
SST	Space Surveillance Telescope
STADAN	Satellite Tracking and Data Acquisition Network
STDN	Spaceflight Tracking and Data Network
STEM	Science, Technology, Education and Mathematics (in education)
STEP	Solar--Terrestrial Energy Program
STS	Space Transportation System (The Space Shuttle); also Space Transportation Systems Ltd
STV	Satellite Test Vehicle
SWS	Space Weather Services
TARS	Turn Around Ranging Station
TDRS	Tracking and Data Relay Satellite
TDRSS	Tracking and Data Relay Satellite System
TERSS	Tasmanian Earth Resources Satellite Station
TIROS	Television and Infra Red Observation Satellite
TOMS	Total Ozone Monitoring System
TTC&M	Telemetry, Tracking, Command and Monitoring
UFO	Unidentified Flying Object
UHF	Ultra High Frequency
UK	United Kingdom
UKUSA	United Kingdom-United States of America Agreement
ULDB	Ultra-Long Duration Balloon
UN	United Nations
UNCOPUOS	United Nations Committee on the Peaceful Uses of Outer Space
US	United States
USSR	Union of Soviet Socialist Republics
V 2	Vergeltungswaffe 2 (Vengeance Weapon 2)
VAST	Viewer Access Satellite Television
VHF	Very High Frequency
VLBI	Very Long Baseline Interferometry
VLF	Very Low Frequency (radio)
VSAT	Very Small Aperture Terminal
VSOP	VLBI Space Observatory Program
VSSEC	Victorian Space Science Education Centre
WA	Western Australia
WASTAC	Western Australian Satellite Technology and Applications Consortium
WGS	Wideband Global Satellite Communications System
WRC	Woomera Range Complex
WRE	Weapons Research Establishment, later Defence Research Centre (DRC) and Defence Science and Technology Organisation (DSTO)
WRECISS	WRE Camera Interception Single Shot
WREDAC	Weapons Research Establishment Digital Automatic Computer
WREROC	WRE Roll Orientation Camera
WRESAT	Weapons Research Establishment Satellite
WTF	Woomera Test Facility
WTR	Woomera Test Range

Timeline of Space Events

Year	Australian Events	International Events
1945	Britain approaches Australia to develop a weapons testing range	World War II ends. Cold War begins
1946	Anglo-Australian joint Project inaugurated Overseas Telecommunications Commission Established	US and USSR begin rocket tests with modified V 2s
1947	Long Range Weapons Establishment created Woomera Rocket Range construction begins	
1949	Australian Defence Scientific Service and Commonwealth Scientific and Industrial Research Organisation established First missile tests at Woomera	
1955	LRWE becomes Weapons Research Establishment	
1957	Minitrack tracking station established Sounding rocket launches and hypersonic research begin at Woomera	International Geophysical Year commences (concludes in December 1958) USSR launches world's first satellite-Sputnik 1
1958	SAO optical tracking station begins operations	US launches first satellites, Explorer 1 and Vanguard 1 NASA established
1959	Australia a founding member of UNCOPUOS	UN Committee on Peaceful Uses of Outer Space established
1960	Muchea and Red Lake Mercury tracking Stations established Island Lagoon deep space tracking station becomes operational	TIROS 1, first weather satellite launched First experimental navigation and surveillance satellites launched
1961	Minitrack and SAO facilities incorporated Into NASA STADAN network Minitrack moved to Island Lagoon	Yuri Gagarin (USSR) first person in space Alan Shepard, first American in space. Mercury program begins
1962	ELDO established	Telstar 1, first commercial communications satellite Mariner 2 reaches Venus. First successful planetary probe
1963	SAO tracking facility moved to Island Lagoon	Valentina Tereshkova (USSR) first woman in space Mercury program ends
1964	Australia is a founding member of INTELSAT First ELDO launch Muchea tracking station closes Carnarvon MSFN tracking station opens Regular reception of satellite weather images begins	INTELSAT established

Timeline of Space Events

Year	Australian Events	International Events
1965	Tidbinbilla DSN tracking station opened Orroral Valley STADAN station operational	Gemini program begins USSR and US achieve first spacewalks INTELSAT 1 first commercial geostationary communications satellite launched France independently launches first satellite Mariner 4 first successful Mars probe
1966	First OTC Satellite Earth Station opens at Carnarvon First direct satellite broadcast between Australia and UK Cooby Creek tracking station opens Red Lake tracking station closes	Luna 9 lands on Moon Lunar Orbiter probes begin mapping Moon's surface Surveyor lunar landing probes commence Gemini program ends
1967	WRESAT, Australia's first satellite launched Australia participates in Our World Honeysuckle Creek Apollo tracking station opens North West Cape Naval Communication Station opens	Our World, first global television broadcast Apollo 1 crew lost in pad fire
1968		Apollo 7 first crewed Apollo mission Apollo 8 first crewed spacecraft to orbit Moon
1969	Australia receives first images from Apollo 11 Operation near Woomera	Apollo 11 makes first crewed landing on the Moon
1970	Joint Defence Facility Nurrungar commences operations at Woomera Joint Defence Space Research Facility, Pine Gap commences operations Australis-OSCAR 5 launched Last ELDO launch in Australia Cooby Creek tracking station closes	Japan launches first satellite China launches first satellite Venera 7 makes first landing on Venus
1971	ELDO withdraws from Australia UK Prospero satellite launched from Woomera	USSR launches world's first space station Salyut 1 Soyuz 11 crew lost in accident
1972	Island Lagoon tracking station closes	Landsat 1 (ERTS), first remote sensing satellite launched Apollo 17 last crewed lunar landing mission Canada establishes world's first domestic satellite network
1973		US Skylab space station launched
1974	Australia declines ESA membership ADSS Abolished; Defence Science & Technology Organisation (DSTO) created Honeysuckle Creek MSFN station transferred to Deep Space Network	Skylab program ends
1975	Australian sounding rocket program ends Carnarvon tracking station closes	Apollo-Soyuz, first US/USSR joint space flight

Timeline of Space Events (Continued)

Year	Australian Events	International Events
1976		Viking 1 makes first successful landing on Mars Indonesia becomes first developing nation to operate a domestic satellite system
1977		First GMS geostationary weather satellite Space Shuttle Enterprise begins flight tests for US Space Shuttle program
1978	WRE formally ceases to exist	Czechoslovakian cosmonaut becomes first non-US/USSR citizen in space
1979	Australia joins INMARSAT UK sounding rocket program ends at Woomera Australian Landsat station opens First Australian SETI program at Parkes	INMARSAT established Voyager 1 and 2 reach Jupiter Skylab re-enters atmosphere, debris lands in Western Australia. Esperance Council issues NASA joke littering fine
1980	Anglo-Australian Joint Project officially ends Starlab project commences Australian Space Industry Symposium	India launches first satellite Voyager 1 reaches Saturn
1981	Aussat established Honeysuckle Creek tracking station closes	First launch of Space Shuttle Columbia Voyager 2 reaches Saturn
1982		US NAVSTAR-GPS navigation system commences operation COSPAS-SARSAT commences operation
1983	Australia rescinds rights to ESA membership Auspace established	US President Reagan proposes Strategic Defence Initiative ('Star Wars') Sally Ride becomes first US woman in space
1984	First National Space Symposium Australian-born Paul Scully-Power flies on Space Shuttle mission STS 41G COSSA established	President Reagan proposes Space Station Freedom
1985	Madigan Report *A Space Policy for Australia* Aussat 1 and 2 launched ARC experiment makes first flight on Space Shuttle mission STS 51C	
1986	Australian Space Board and National Space Program established Cape York Spaceport proposal	Space Shuttle Challenger destroyed USSR Mir space station launched Voyager 2 reaches Uranus Giotto mission visits Comet Halley
1987	Australian Space Office established Aussat 3 launched	

Timeline of Space Events

Year	Australian Events	International Events
1988	First Ausroc student rocket launched ARC experiment makes second flight on STS 26	US Space Shuttle flight resume after Challenger disaster Cosmonauts spend full year in space Israel launches first satellite
1989	Cole Report on Australian space science Australia joins COSPAS-SARSAT network	Voyager 2 reaches Neptune Hipparchos space telescope launched
1990	Balanced National Space Program introduced	Hubble Space Telescope launched Magellan space probe begins radar mapping of Venus Hiten, first Japanese Moon probe launched
1991	ERS 1 launched with ATSR	
1992	Endeavour space telescope flies on STS 42 Space Shuttle mission Curtis and BIE Reports review NSP Aussat privatised-becomes Optus OTC merged with Telecom Australia Australian Space Industry Chamber of Commerce established	International Space Year
1993	ASB replaced by Australian Space Council First successful scramjet engine test	Space Station Freedom program become International Space Station with Russia as partner
1995	Interdepartmental Committee on International Space review of NSP Project Phoenix SETI search Parkes Endeavour space telescope second flight on STS 67 ERS 2 launched with ATSR 2	Galileo space probe arrives at Jupiter First Shuttle-Mir mission
1996	ASO, ASC and NSP terminated FedSat announced, COSSA restructured Australian-born astronaut Andy Thomas makes first spaceflight on STS 77	
1997		Mars Pathfinder mission lands on Mars
1998	Australian Space Act, SLASO established CRCSS established for FedSat program Andy Thomas resumes Australian citizenship	Andy Thomas becomes last US astronaut on Mir in Shuttle-Mir program First components of ISS launched XMM-Newton space telescope launched
2000		First crew occupies ISS
2001	First HyShot launch Australian Centre for Astrobiology established Andy Thomas third flight on STS 102	Intelsat privatised Mir space station de-orbited
2002	FedSat launched Envisat launched with AATSR	

Timeline of Space Events (Continued)

Year	Australian Events	International Events
2003	Australian Government Space Engagement Policy Framework released	Space Shuttle Columbia destroyed China launches first astronaut, Yang Li Wei Japan's Hayabusa asteroid mission launched Mars Express first ESA Mars mission First cubesats launched
2004		Mars Exploration Rovers arrive on Mars Cassini mission arrives at Saturn
2005	CRCSS closed *Space: a Priority for Australia* released Andy Thomas fourth flight on STS 114	ESA Huygens probe lands on Titan Space Shuttle flights resume after Columbia accident
2006	AITC opens at Mount Stromlo, Canberra	Venus Express first ESA Venus probe
2007	FedSat ceases operation	Peggy Whitson becomes first female commander of ISS Chang'e 1, first Chinese lunar mission
2008	Senate Standing Committee on Economics' *Lost in Space?* Report	Chandrayaan, first Indian lunar mission launched SpaceX Falcon 1 first privately-developed launch vehicle to reach Earth orbit
2009	ASRP commences First HIFiRE flight	US radio host pays joke Skylab fine (1979)
2010	Japan's Hayabusa sample return capsule lands at Woomera	
2011		Space Shuttle program ends China launches Tiangong 1 space station
2012		Curiosity rover lands on Mars
2013	Australia's Satellite Utilisation Policy released ASRP terminated	Chang'e 3, first Chinese lunar rover Mangalyaan first Indian Mars mission launched Voyager 1 leaves Solar System
2014		ESA Rosetta mission arrives at comet 67P
2015		New Horizons arrives at Pluto US Year in Space mission on ISS begins First successful landing of SpaceX re-usable rocket decent stage
2016		Juno mission arrives at Jupiter
2017	SIAA Space White Paper First Australian cubesats launched	

About the Author

Kerrie Dougherty

Kerrie Dougherty is a freelance curator, historian, educator and writer, with more than thirty years' experience in the space field. Formerly Curator of Space Technology at the Powerhouse Museum, Sydney, Kerrie co-authored *Space Australia*, the first popular history of Australian space activities, on which this book is based. A member of the Faculty of the International Space University, Kerrie lectures in Space Humanities in the ISU's Space Studies Program and Southern Hemisphere Space Studies program. She is also an elected Member of the International Academy of Astronautics and serves on international committees on the 'history of astronautics', 'space education and outreach' and 'space and museums'. Kerrie has been the recipient of an Australian Space Pioneer Award from the National Space Society of Australia and was the 2015 winner of the Sacknoff Prize for Space History.

Acknowledgements

Australia in Space is a fully revised and expanded second edition of *Space Australia*, originally published in 1993 by the Powerhouse Museum, Sydney, as part of the Museum of Applied Arts and Sciences. SIAA gratefully acknowledges the Museum's permission to produce this new edition. Matthew L James, co-author of *Space Australia*, is also acknowledged for his significant contribution to the original edition.

The support of the Defence Science and Technology Group in providing historic photographs was invaluable and special thanks are due to Ms Julie Bebbington, Manager of Defence Science Communications, Dr Ruth Donovan, Research Science Historian, and Mr Brian Holland, Defence Records Management (SA) for their kind assistance.

The website www.honeysucklecreek.net and its webmaster/curator Colin Mackellar are gratefully acknowledged for providing access to many photographs documenting the history of space tracking in Australia. Website contributors Guntis Berzins, Hamish Lindsay and Ed von Renouard are specifically thanked for permission to use their images featured on the website.

Dr Andrew SW Thomas, NASA astronaut and first Australian citizen in space, is also thanked for providing the Foreword for this edition.

The author and publishers would also like to acknowledge and thank the following agencies, corporations and individuals for providing images, information and support:

Advanced Instrumentation Technology Centre, Research School of Astronomy & Astrophysics, ANU (Dr Naomi Mathers and Mr Mike Petkovic)
Astronomical Society of Western Australia
Auspace, part of the Nova Group
Australia Postal Corporation
Australian Centre for Space Engineering Research, University of New South Wales (especially Professor Andrew Dempster and Ms Cheryl Brown)
Mrs Ann Cairns and Professor Iver Cairns, School of Physics, University of Sydney
Centre for Hypersonics, University of Queensland
Commonwealth Scientific and Industrial Research Organisation (CSIRO)
Department of Defence
Department of Industry, Innovation and Science
DLR-Archiv Köln
Mr David Dougherty
European Space Agency (ESA) ESA images on pages 108 and 125 are released under Creative Commons Attribution 3.0 IGO. https://creativecommons.org/licenses/by-sa/3.0/
Geoscience Australia Geoscience Australia images are released under the Creative Commons Attribution 4.0 International Licence. http://creativecommons.org/licenses/by/4.0/legalcode
Dr Brett Gooden
Mr Warwick Holmes
Japan Meteorological Agency (GMS images available at http://www.jma.go.jp/en/gms/)
Lunar and Planetary Institute
Dr Owen Mace
Mt Stromlo Observatory, Research School of Astronomy & Astrophysics, ANU
Mr Glen Nagle, Canberra Deep Space Communications Complex
National Academy of Sciences, United States
National Aeronautics and Space Administration (NASA)
National Oceanic and Atmospheric Administration (NOAA)
NBN Co.
Optus Satellite
Queensland Police Museum
Dr John Sarkissian, CSIRO Parkes Radio Telescope
Dr Ravi Sood, University of NSW, ADFA campus
State Library of New South Wales
Status International, Sydney
Telstra Corporation Ltd
Mr Richard Tonkin
US Geological Survey, US Department of the Interior
US National Executive Committee for Space-Based Positioning, Navigation and Timing
Mr Martin Walker

INDEX

A

ADSS, 14, 15, 16, 17, 18, 79, 165, 166, 178, 183.
Advanced Along Track Scanning Radiometer, 124, 125, 142, 144, 178.
Advanced Earth Observation Satellite, 125, 178.
Advanced Research Projects Agency (ARPA), 75, 88.
Aero High rocket, 38, 39.
Aero Mach (hypersonic research rocket), 77.
Aero space planes, 87, 129.
Aerobee (sounding rocket), 76.
AETHER (Advanced Equipment to Handle ERS), 144, 145, 178.
Afghanistan, 21, 85.
Africa, 53.
Aggregation of Red Cells (ARC) experiment, 126, 178.
Aldrin, 'Buzz' (astronaut), 64.
Alesi Technologies, 87.
ALFLEX (Automatic Landing Flight Experiment, 79, 178.
Alice Springs, 85, 96, 104, 105, 110, 118.
Along Track Scanning Radiometer (ATSR), 123, 124, 125, 142, 146, 151, 170, 178, 185.
Amateur space projects, 22, 23, 147.
American Interplanetary Society, 3.
AMSAT (Amateur Radio Satellite Corporation), 48, 49, 178.
Anglo-Australian Joint Project, 11, 25, 39, 76, 78, 164, 165, 182.
Anik A1 (satellite), 95.
Antarctic bases, 94, 100.
Antarctica, 101, 106, 115, 134, 143, 157.
Antennas, 34, 69, 70.
Anti-ballistic missile (ABM), 75, 78, 178.
AOTC (Australian and Overseas Telecommunications Corporation), 93, 96, 178
Apollo 11, iv, v, 64, 65, 70, 71, 120, 121, 183.
Apollo Program, 52, 55, 56, 58, 61, 62, 63, 64, 65, 66, 75, 76, 117, 119, 120, 121, 134, 135, 178.
Apollo-Soyuz Test Project, 55, 135, 183.
Application Technology Satellite (ATS), 66, 95, 100, 178.
Applications satellite, 55, 66, 92.
Applications Technology Satellite (ATS), ATS 1, ATS 2, 66, 100.
AQUA (satellite), 102.
Ariane (rocket), 68, 95, 97, 166.
Ariane program, 68, 95, 97, 166.
Arms control, 85, 86.
Armstrong, Neil, 63, 64, 65.
ASERA (Australian Space Engineering Research Association Ltd), 147, 178.
Asia-Pacific region, 79, 98, 152, 173, 178.
Astris rocket, 31.
Astronauts, 56, 61, 62, 63, 64, 117, 119, 120, 134, 135, 136.
Astronomy, 22, 36, 54, 57, 68, 94, 113, 115, 116, 117, 118, 119, 121, 123, 125, 127, 154, 155, 157, 167, 170.
Astrotech Space Operations, 87.
Atmospheric Pressure Censor, 125, 178.
Atmospheric studies, 45, 102, 118, 136.
Atock, James Kenneth ('Ken'), 5.
ATS-1 satellite, 66.
Auspace Ltd, 121, 122, 124, 125, 131, 142, 143, 147, 151, 152, 155, 184.
Ausroc program, 40, 147, 178, 185.
Aussat, 66, 94, 95, 96, 97, 137, 141, 142, 143, 166, 167, 170, 184, 185.
Australia Day, Expo 112, 113.
Australian Academy of Science, 164, 170.
Australian Academy of Technological Science, 144, 168.
Australian and Overseas Telecommunications Corporation (AOTC), 96, 170, 178.
Australian Centre for Remote Sensing (ACRES), 104, 105, 123, 128, 142, 144, 170, 178.
Australian Centre for Signal Processing, 145.
Australian Centre for Space Engineering Research (ACSER), 81, 160, 178.
Australian Defence Force Academy (ADFA), 88, 118, 178.
Australian Defence Satellite Communication Station (Kojarena), 86.
Australian Defence Scientific Service (ADSS), 14, 15, 16, 17, 18, 79, 140, 165, 166, 178, 183.
Australian Geospatial-Intelligence Organisation (AGO), 86, 178.

Australian Hypersonics Initiative (AHI), 88.
Australian Launch Vehicle (ALV), 147, 178.
Australian National University (ANU), 24, 87, 114, 120, 121, 122, 128, 129, 130, 131, 142, 155, 157, 178.
Australian Research Council (ARC), 43, 87.
Australian Rocket Society (ARS), 1, 3, 4, 5, 178.
Australian Signals Directorate (ASD), 85, 86, 178.
Australian Space Board (ASB), 144, 146, 168, 169, 170, 171, 172, 178, 185.
Australian Space Industry Development Strategy, 145, 169.
Australian Space Industry Symposium, 121, 141, 185.
Australian Space Office, 122, 124, 126, 142, 145, 146, 167, 168, 169, 170, 172, 178.
Australian Space Research Agency (ASRA), 166, 178.
Australian Space Research Institute (ASRI), 147, 178.
Australian Spaceport Group (ASG), 149, 178.
Australian Surveying and Land Information Group (AUSLIG), 104, 170, 178.
Australis (Australia-Oscar 5), 47, 48, 49.

B

Baker-Nunn camera/Baker-Nunn telescope, 19, 20, 22, 51, 52, 55.
Ballima (tracking dish), (DSS 43), 59.
Balloons, 1, 99, 118, 179.
Belgium, 31, 160.
Belrose, 96.
Bendigo (Vic), 86, 98.
Biarri Point, 81, 82.
Biarri, 81.
BIOS (biological satellite), 62, 178.
Black Arrow rocket, 34, 76, 78.
Black Knight rocket, 24, 30, 31, 35, 74, 76, 87, 164.
Blue Streak rocket, 29, 30, 31, 32, 61, 74, 75, 78, 164.
Boeing, 88, 142.
Bourke (NSW), 99.
Boxer rocket, 2.
Brisbane, 3, 4, 5, 66, 100.
British Aerospace Australia (see also BAe Systems), 81, 124, 130, 140, 142, 144, 178.
British Aerospace, 124, 129, 140, 142, 144, 147, 178.
Broken Hill (NSW), 98, 118.
Buccaneer (satellite), 82.
Bureau of Industry Economics, 146, 171, 178.
Bureau of Meteorology, 104, 169, 170, 178.
Bureau of Mineral Resources, 104, 170.

C

Cameras (WRE)- WREROC, WRECISS, 17, 18, 32.
Canada, 7, 45, 66, 81, 85, 95, 110, 112, 113, 121, 134, 142, 164, 167, 168, 183.
Canberra Deep Space Communications Complex (Tidbinbilla tracking station), 54, 60, 125, 169, 178.
Canberra, 22, 43, 46, 51, 54, 55, 57, 58, 63, 66, 79, 80, 82, 86, 88, 104, 118, 25, 152, 155.
Capcom (Capsule Communicator), 61, 63, 178.
Cape Canaveral, 14, 16, 50, 53, 61, 67, 81.
Cape York International Spacelaunch Ltd (CYISL), 150.
Cape York Space Agency (CYSA), 149.
Cape York Space Port, 148, 150, 171, 184.
Carnarvon (WA), 37, 38, 44, 61, 62, 63, 66, 68, 93, 94, 99, 112.
Carnarvon Satellite Earth Station, 183.
Carnarvon tracking station, 37, 38, 44, 53, 61, 62, 63, 64, 66, 68, 93, 94, 99, 111, 112, 182, 183.
Carver, John, 25, 38, 41, 43, 46, 116.
Cassini (spacecraft), 61, 71, 186.
Ceduna (SA) (OTC Satellite Earth Station), 99.
Ceduna satellite ground station, 94, 99, 128.
Central Australian Aboriginal Media Association, 96.
Central Intelligence Agency (CIA), 178.
Centre for Hypersonics, University of Queensland, 87, 88, 89, 129, 130, 158.
Centrifuge, 6.
Challenger space shuttle, 55, 122, 126, 135, 184.
Chapman, Philip K, 34, 134.
Chevaline, 78.
Chifley Labor Government, 11.
Chifley, Ben, Prime Minister of Australia, 11.
China, 2, 69, 85, 101, 102, 109, 183, 186.
Christmas Island, 11, 98, 148, 151.
Cluster (spacecraft), 68, 69.
CNES (French space agency), 97, 119, 178.
Cocos (Keeling) Islands, 86.
Cocos Islands, 98.
Cold War, 11, 19, 25, 30, 74, 82, 84, 85, 86, 107, 108, 109, 110, 163, 181.

Cole Report, 185.
Columbia Command Module (Apollo 11), 64.
Columbia space shuttle, 135, 137, 184, 186.
Comet Haley, 67, 71, 86.
Committee on the Peaceful Uses of Outer Space (COPUOS) (also UNCOPUOS), 25, 178, 182.
Commonwealth Scientific and Industrial Research Organisation (CSIRO), 25, 50, 54, 60, 64, 66, 69, 70, 104, 113, 116, 118, 123, 124, 125, 126, 127, 142, 143, 144, 145, 148, 151, 152, 153, 160, 164, 167, 168, 170, 172, 178, 179.
Congreve rocket, 2.
Cooby Creek tracking station, 66, 90, 95, 113, 181.
Corella rocket, 39.
Cosmic ray, 57, 117, 118, 119.
Cosmonauts, 134, 185.
Cospas-Sarsat program, 110, 179, 184.
Cotton Aero-dynamic Anti-G (CAAG) suit, 6, 178.
Cotton, Francis Stanley ('Frank'), 6, 7.
Crib Point Satellite Earth Station (Vic), 102.
CSIRO Office of Space Science and Applications COSSA), 105, 123, 124, 125, 126, 144, 151, 167, 168, 169, 170.
CSIRO Upper Atmosphere Section, 25, 116.
Cubesat/cubesats, 81, 82, 83, 157, 158, 159, 160, 161, 177, 179.
Curiosity (Mars rover), 60, 71, 86.
Cyclones, 99, 100.

D

Darwin, 34, 66, 86, 100, 151.
Davis Cup, 66, 112.
Deakin NASCOM Centre, 54.
Deep Space Antenna 1 (DSA 1), 68, 179.
Deep Space Instrumentaton Facility, 55, 56, 179.
Deep Space Network (DSN), 51, 52, 56, 66, 71, 127, 179.
Defence Evaluation and Research Agency (DERA) (see also QinetiQ), 87, 88.
Defence Imagery and Geospatial Organisation (DIGO), 86, 179.
Defence Research Centre (DRC, DRCS), 79, 179.
Defence Research Laboratories, 14.
Defence Satellite Communications System (DSCS), 84.
Defence Science and Technolgy Organisation (DSTO), 79, 81, 83, 87, 88, 130, 151, 179.
Defence Science and Technology Group (DST Group), 9, 13, 16, 20, 24, 27, 28, 30, 32, 35, 36, 38, 39, 40, 43, 45, 74, 81, 82, 83, 88, 130, 140, 165, 179.
Defence Signals Directorate, 85, 179.
Defence Support Program (satellites), 84.
Defence White Paper, 89.
Defense Satellite Communications System (DSCS), 84, 178.
Delta DOR (Delta Differential One-Way Ranging), 69.
Delta rocket, 69, 161.
Dennett, John, 2.
Department of Civil Aviation (DCA), 66, 179.
Department of Defence (Australia), 84, 87, 165, 173.
Department of Industry Science and Resources, 87.
Department of Supply, 40, 48, 53, 54, 79, 113, 118, 165, 166, 178.
Dintenfass, Leopold, 126.
Disaster monitoring, 102, 106, 157.
Discovery space shuttle, 95, 122, 136.
DISIMP software, 143, 179.
Dixon, Frank, 18.
DLR (German Aerospace Centre), 87, 117, 142, 179.
Dongara tracking station (WA), 66, 68.
Doppler Orbitography and Radiopositioning Integrated by Satellite (DORIS)
Down Under Comes Up Live (satellite broadcast), 94, 111, 112.
Dropsonde, 37, 39.
DSS 34, 60.
DSS 35, 60.
DSS 36, 60.
DSS 41, 56, 57, 58.
DSS 42, 58.
DSS 43, 59.
DSS 44, 66.
DSS 45, 60, 66.
DSS 46, 60.

E

Early Bird (satellite), 92.
Early warning, 74, 80, 82, 84, 85.
Earth Resources Technology Technology Satellite (ERTS-1), 103, 179.

Earth station, 67, 94, 102, 111.
Earth, 18, 23, 34, 37, 41, 43, 44, 47, 48, 49, 50, 51, 52, 53, 54, 55, 56, 57, 58, 59, 60, 61, 63, 66, 67, 68, 71, 73, 83, 91, 92, 93, 94, 99, 100, 101, 102, 103, 104, 105, 106, 107, 108, 109, 110, 111, 116, 120, 121, 124, 125, 127, 128, 131, 132, 135, 136, 146, 150, 151, 152, 156, 158, 160, 172, 178, 179180, 181, 183, 186.
ECHELON, 85, 86.
Education, 25, 66, 96, 132, 147, 151, 153, 154, 156, 158, 159, 161, 170, 176, 177.
ELDO, 26, 28, 30, 31, 32, 33, 34, 35, 41, 44, 67, 74, 75, 78, 140, 147, 165, 166, 178.
Electro Optic Systems (EOS), 83, 179.
Electronic intelligence (Elint), 83, 179,
Emergency Position Indicating Radio Beacon (EPIRB), 110, 179.
Emu Field, 14.
Endeavour space shuttle, 122, 123, 136, 147.
Endeavour telescope, 122, 123, 143, 145, 146, 185.
Environment, 11, 18, 23, 30, 35, 56, 61, 68, 80, 86, 97, 99, 102, 103, 107, 116, 123, 126, 155, 159, 160.
Environmental data, 86, 104, 125, 146.
Environmental monitoring, 102, 104, 106, 125, 146, 168, 171.
Environmental Science Services Administration (ESSA), 63.
Envisat, 104, 108, 124, 142, 179, 185.
Estrack, 67, 68.
Etherington Optical, 18.
Eumetsat, 68, 179.
Europa rocket, 30, 31, 32, 33, 75.
European Launcher Development Organisation (ELDO), 26, 29, 30, 31, 32, 33, 34, 35, 41, 44, 67, 74, 75, 78, 140, 147, 165, 166, 178, 179, 182, 183.
European Remote Sensing (ERS) satellites, 107, 168, 179.
European Space Agency, 122, 124, 166, 179.
European Space Operations Centre (ESOC), 67, 179.
European Space Research Organization (ESRO), 33, 179.
Exmouth (WA), 84.
Expert Panel, 146, 150, 171.
Explorer, 21, 22, 41, 120.

F

Fairey Aviation Company of Australasia Pty Ltd, 17, 23, 140.
Falling sphere, experiments, 37.
Falkenberg, Brian, 3.
Falstaff (hypersonic research rocket), 76.
Fast Delivery Processor (FDP), 144, 179.
Feng Yun (satellite), 101, 102.
First National Space Symposium, 168, 184.
Five Eyes, 81, 85.
Flight Surgeon, 61, 62.
Forbes-Martyn, David, 24.
France, 31, 34, 35, 45, 68, 110, 183.
Fraser Government, 95, 104, 141, 167.
Freedom space station, 121, 185.
Fuse/Lyman ultraviolet telescope, 122, 142, 145, 179.

G

Gagarin, Yuri, 61, 134, 182.
Gaia (spacecraft), 68, 69.
Galileo, space probe, 71, 109, 185.
Gascan ('Get Away Special' canisters), 122, 141, 179.
Geeveston (Tas), 99.
Gemini program, v, 61, 62, 63, 75, 93, 183.
George V, King, I, 4.
George VI, King, 5.
Geostationary Meteorological Satellite (GMS), 101, 102, 143, 148, 179.
Geostationary Meteorological Satellite (see also GMS Himawari, Himawari) 101, 102.
Geostationary orbit, 34, 92, 101, 103, 110, 146, 179.
Geraldton, 86, 99.
Germany, 31, 34, 67, 69, 118, 130.
Gile Meteorological Station, 26.
Giotto space probe, 67, 71, 184.
Glenn, John (astronaut), v, 50, 61.
Global Atmospheric Methane Senor (GAMS), 125, 148, 179.
Global village, 99.
Glonass system, 108.
Gnangara satellite ground station, 68, 94, 98, 128.
Goddard Space Flight Centre, 52, 54, 117.
Goddard, Robert, 3.
Google Earth, 83, 109.
Gove tracking station, 32, 34, 44.

GPS, 80, 82, 107, 108, 109, 110, 127, 142, 146, 152, 157, 160, 161, 179.
Ground station, 63, 74, 82, 83, 85, 86, 93, 94, 95, 96, 97, 98, 99, 104, 110, 140, 143, 167, 171, 174.
Gulf War, 84.

H

Hawke Government, 79, 114, 168.
Healesville (Vic) (OTC Satellite Earth Station), 94.
Helios (spacecraft), 66, 117.
Herschel (spacecraft), 68.
High Altitude Balloon (HAB), 48, 118.
High Altitude Density (HAD), 37, 179.
High Altitude Research Project (HARP) (sounding rocket), 22.
High Altitude Sounding Projectile (HASP), 39, 179.
High Frequency (HF), 82, 179.
Hipparchos (spacecraft), 68, 185.
HMAS Cerberus Naval Training BaseHypersonic International Flight Research Experimentation (HIFiRE), 40, 88, 89, 130, 179, 186.
Homestead and Community Broadcast Satellite Service (HACBSS), 66, 96, 179.
Honeysuckle Creek (also DSS 46), 50, 53, 55, 57, 58, 60, 62, 63, 64, 65, 66, 90, 94, 111, 112, 163, 164, 183, 184.
HOPE spaceplane, 79, 151, 179.
Howard Government, 79, 80, 123, 125, 136, 146, 151, 172, 173, 174.
Hubble Space Telescope, 92, 121, 185.
Hustler, Frederick, 2.
Huygens (spacecraft), 69, 71, 186.
HyCAUSE (Hypersonic Collaborative Australian/ United States Experiment), 88, 180.
Hyperion (hypersonic research rocket), 78.
Hypersonic research, 40, 75, 76, 77, 78, 80, 87, 88, 128, 129, 130, 158, 165, 179.
Hypersonic Research Vehicle (HRV), 77, 78.
hypersonic shock tunnel, 87, 128.
HyShot, 40, 87, 88, 89, 130, 185.

I

ICBMs, 30, 73, 83, 84.
Ikara (anti-submarine weapon), 15, 18.
Imparja Television, 96.

INMARSAT (International Maritime Satellite Organisation), 69, 93, 94, 97, 143, 180, 184.
INTELSAT I, II, III (satellites), 63, 90, 92, 93, 94, 97, 111, 143, 180.
International Mobile Satellite Organization (IMSO), 97, 180.
International Space Station (ISS), 82, 121, 136, 137, 150, 159, 160, 174, 180.
Interplanetary Monitoring Platform, 62, 180.
Iraq, 84.

J

Jabiru (hypersonic research rocket), 76, 77, 78.
Jet Propulsion Laboratory, 52, 54, 136, 180.
Jindalee Operational Radar Network (JORN), 72, 82, 180.
Jindivik (target aircraft), 14, 18.
Joint Defence Facility Nurrungar, 73, 79, 84, 85.
Joint Defence Facility, Pine Gap, 84, 85, 86, 183.
Joint Defence Space Communications Station / Joint Defence Facility Nurrungar, 79, 84.
Joint Defence Space Research Facility, 85, 183.
Jupiter, 57, 71, 184, 186.

K

Kalgoorlie (WA), 99.
Kojarena (WA), 86.
Kokatha (Aboriginal people), 27.
Kourou (French Guiana launchsite), 34, 35, 67, 68.

L

Lake Torrens, 11.
laser ranging, 66, 67, 81, 157, 180, 181.
Lifting body/bodies, 76.
LISA Pathfinder (spacecraft), 68.
Lobster (Australian rocket motor), 77.
Long March 2e (rocket), 97.
Low Earth orbit (LEO), 92, 103, 110, 146, 180.
Lunar Orbiter (spacecraft), 56, 57, 58, 183.
Luxfer, Australia, 87.

M

McLuhan, Marshall, 99.
Macquarie Island, 98.
Magellan (spacecraft), 51, 185.
Magellan (giant telescope), 155.

Malkara (anti-tank weapon), 15, 17, 18, 38.
Mangalyaan (spacecraft), 60, 186.
Manned Space Flight Network (MSFN) (see also Mercury Space Flight Network), 51, 52, 61, 62, 180.
Marecs (spacecraft), 67.
Mariner (spacecraft), 56, 57, 58, 70, 182.
Marloo (rocket motor), 18, 38, 77.
Mars, 56, 57, 58, 59, 60, 66, 68, 70, 71, 132, 156, 159.
Mars Express (spaeccraft), 69.
Martyn, David Forbes, 25, 116.
Melbourne, 5, 22, 48, 66, 94, 100, 101, 113, 118, 134, 147, 157.
Melbourne University, 47, 48, 118, 120, 159, 180.
Melbourne University Astronautical Society, 47, 180.
Melbourne University Radio Club, 47.
Menzies Liberal Government/Menzies Government, 14, 164.
Mercury (Ken Atock rocket), 5, 52, 61, 62, 182.
Mercury (planet), v, 68.
Merritt Island Launch Area (MILA), 20, 180.
Meteosat (spacecraft), 67.
Mildura (Victoria), 18, 118.
Mobile Laser Station (MOBLAS) 5, 66, 180.
Monash University, 147.
Moon, the, iv, v, 7, 25, 36, 52, 55, 56, 58, 61, 62, 63, 64, 65, 66, 67, 70, 71, 119, 120, 121, 134, 183, 185.
Moonwatch, 20, 21, 22, 23, 52.
Moree (NSW) (OTC Satellite Earth Station), 94, 111.
Morrison, Noel, 3, 4.
MOS 1 (spacecraft), 68, 105, 180.
Mount Stromlo, 22, 81, 114, 121, 131, 139, 142, 155, 157, 186, 188.
Multi-functional Transport Satellite (MTSAT), 101, 180.

N

Namuru GPS receivers, 81.
NASCOM (NASA Communications), 54, 180.
National Aerospace Laboratories (NAL), 87.
National Broadband Network (NBN), 98, 180.
National Committee for the International Geophysical Year (Australian Academy of Science), 21.
National Indigenous Television (NITV), 96.
National Oceanic and Atmospheric Administration (NOAA), 101, 180
National Reconnaissance Office (NRO), 85, 180.
National Security Agency (NSA), 85, 180.
Naval Research Laboratory (NRL), 19, 52, 180.
Neptune, 58, 71, 127, 185.
New Horizons (spacecraft), 60, 186.
New Norcia (European Space Agency tracking station), 51, 68, 69, 71.
New Zealand, 81, 85, 96, 98, 103, 106, 120.
NEXST 1 (Japanese supersonic transport aircraft), 41, 79, 89, 180.
Nimbus (satellite), 47, 100.
Norfolk Island, 98.
North West Cape Naval Communications Station, 84, 183.
NQEA Australia, 87.
nuclear weapons, 14, 30, 84.
Nulka (active missile decoy), 18.
Nurrungar, 73, 79, 84, 85, 183.

O

Oberon, 76.
Opportunity (Mars rover), 71.
Optus, 151, 156, 157, 169, 171, 185.
Optical Group, ADSS, 17, 18.
Optus Aurora (television service), 96.
Optus B1, 97.
Optus B2, 97.
Optus C1 (satellite), 81, 97.
Optus Communications, 96.
Optus 10, 98.
Optus 13, 142.
Orbiting Geophysical Observatory, 62, 63, 66.
Orion (rocket), 1, 4, 87.
OTC (Overseas Telecommunications Commission/Overseas Telecommunications Corporation), 21, 22, 54, 63, 64, 67, 68, 93, 94, 96, 99, 111, 113, 142, 143, 167, 169, 180, 183, 185.
Our World (global satellite broadcast), 66, 113, 183.
Outer Space Treaty (UN), 25, 84.
Oxford Falls (NSW) (OTC Satellite Earth Station), 94, 98.

P

Packer, Kerry, 95.
Palapa (satellite), 95.
Papua New Guinea, 34, 96, 134, 151.

Parkes Radio Telescope, 56, 59, 64, 65, 66, 67, 69, 70, 71, 113, 127, 133.
Perth (WA), v, 21, 22, 53, 61, 64, 66, 68, 94, 96, 97, 104, 111, 112, 133, 139.
Perth International Telecommunications Centre, 68, 94, 98, 139.
Phytotron (CSIRO), 66, 113.
Pine Gap, 84, 85, 86, 181.
Pioneer (spaeccraft), 56, 57, 58, 69, 117.
Pioneer Venus (spacecraft), 66.
Planck (spacecraft), 68.
Pluto, 60, 186.
Polar orbit, 41, 92, 101, 102, 103, 147.
Polaris missile, 78.
Postmaster General's Department (PMG), 54, 66, 95, 113, 180.
Prime A stations (Moonwatch), 23.
Prime Minister Harold Holt, 66, 112.
Project SPARTA, 41, 42, 43, 72, 75, 76, 130.
Propulsion Research Laboratory, 18, 52, 136.

Q

Quasar 0237-23, 66.
Queensland Air Mail Society, 3.
Queensland Government, 87, 148, 150.

R

R.T. (ARS rocket series), 4, 5.
RAAF Aircraft Research and Development Unit (ARDU), 79, 80, 87, 178.
Red Lake (tracking station), 53, 62, 182, 183.
Redstone (rocket), 41, 44, 46, 72, 76.
Research Projects Agency (DARPA), 75, 88.
Rocket mail, 3, 4, 5.
Rockingham (WA), 61.
Rockoon, 23, 24, 180.
Rohini (spacecraft), 68.
Roma (Qld), 99, 158.
Rosetta (spacecraft), 68, 186.
Royal Aircraft Establishment (RAE), 77, 180.
Royal Australian Air Force (RAAF), 6, 14, 79, 80, 84, 119, 147, 178, 180.

S

Satellite geodesy, 83.
Saturn, 65, 71.
Schmeidl, Friedrich, 3.
SCRAMJET (Supersonic Combustion Ramjet), iv, 87, 88, 89, 129, 130, 131, 157, 158, 181.
Scud missile/s, 84.
Sentinel (spacecraft), 68.
Seoul National University, 87.
Shoal Bay (NT), 86.
Signals intelligence (Sigint), 83, 85, 181.
Sky Muster (satellite), 98.
Skylark (sounding rocket), 9, 23, 24, 28, 36, 37, 38, 40, 43, 45, 116, 117, 118, 165.
SMART 1 (spacecraft), 68.
Smithsonian Astrophysical Observatory (SAO), 19, 20, 21, 22, 52, 181.
Solar Particle Alert Network (SPAN), 63, 181.
Sounding rockets, 18, 23, 24, 29, 35, 36, 37, 38, 39, 40, 41, 44, 74, 75, 76, 80, 99, 116, 147, 158, 165.
South Australian Government, 2, 87, 147.
South-East Asia, 85, 97, 101, 168.
Soyuz (rocket), 55, 68, 135, 150, 151, 183.
Space Co-operation Agreement, 53, 54.
Space Projects Branch, Department of Science, 79.
Space Shuttle, v, 66, 95, 114, 122, 123, 126, 129, 135, 136, 137, 181.
Space Situational Awareness (SSA), 83, 84, 181.
Space Surveillance Network, 84, 85.
Space tourism, 79, 80.
Space Tracking and Data Acquisition Network (STADAN), 43, 51, 52, 53, 54, 55, 57, 62, 66, 67, 181, 182, 183 (see also Space flight Tracking and Data Network, or STDN), 54, 60, 180.
Spaceflight Movement, 3.
Sputnik 1, 19, 20, 21, 84, 92, 182.
SS Maheno, 3.
Stamp, WRESAT, 46.
Stanley (Hong Kong), 98.
Stonechat (rocket motor), 78.
Sugar scoop (satellite antenna), 94, 112.
Suomi NPP (satellite), 91, 100, 102.
surveillance satellites/ spy satellites, 82, 84, 103, 182.
Surveillance Telescope (SST), 84, 181.
Surveyor (spacecraft), 58, 183.

Swigert, Jack (astronaut), 65.
Syria, 85.

T

Talos-Castor (sounding rocket), 88.
Telecom Australia (see also PMG and Telstra), 95, 96.
Teleeducation, 99.
Telehealth, 99.
Teleports, 98.
Telstar, 92.
Telstra, 96, 97, 98, 139, 143, 169.
TERRA (satellite), 102.
TIROS (Television and Infra-Red Observation Satellite), 47, 48, 99, 100, 181, 182.
TIROS 1, 99, 182.
Titan (moon of Saturn), 71.
Tracking and Data Relay Satellite System (TDRSS), 54, 55, 181.
Tsiolkovski, Konstantin E, 3.
Turn Around Ranging Station (TARS), 101, 102, 181.

U

UKUSA (agreement), 85, 181.
University College London, 36.
University of Adelaide, 22, 36, 38, 41, 43, 46, 116, 117, 136, 161.
University of Leicester, 36.
University of Melbourne, 47, 48, 49, 118, 121, 159.
University of New South Wales, 55, 81, 82, 87, 130, 132, 156, 161, 175.
University of Queensland, 87, 88, 89, 128, 129, 130, 157, 158, 188.
University of South Australia, 145, 151, 156, 157, 159, 161.
University of Sydney, 110, 126, 134, 135, 156, 161.
University of Tasmania, 36, 55, 67, 94, 104, 117, 118.
Unmanned Aerial Vehicle/s (see also drone/s), 80, 161.
Uranus, 59, 60, 71, 127.
US Air Force (USAF), 47, 84.
US Air Force Office of Scientific Research (AFOSR), 87.
US Air Force Research Laboratory (AFRL), 88, 178.

V

van Allen radiation belts, 92.

Vanguard (satellite), 18, 19, 20, 22, 182.
Vega (rocket), 68.
Venus, 51, 56, 58, 66, 69, 70, 182.
Very Long Baseline Interferometry (VLBI), 67, 127, 181.
Viewer Access Satellite Television (VAST), 96, 181.
Viking (spacecraft), 59, 60, 66.
von Karman Award for International Cooperation in Aeronautics, 89.
Vostok, 61.
Voyager (spacecraft), 59, 60, 66, 71, 127, 184, 185, 186.

W

Waroona (WA), 99.
Weapons Research Establishment, 12, 14, 16, 20, 53, 79, 179, 181.
Weemala (tracking dish) (DSS 42), 57, 59.
Wideband Global SATCOM network (WGS), 81, 86, 181, 182.
Wolumla (NSW), 98, 99.
Woomera Immigration Reception and Processing Centre, 80.
Woomera Launch Area, 8, 32, 44, 76.
Woomera Prohibited Area, 11, 27, 79.
Woomera Range Complex (WRC), 80, 181.
Woomera Test Facility (WTF), 11, 79, 181.
Woomera Test Range (WTR), 5, 9, 11, 12, 14, 23, 24, 26, 27, 30, 31, 34, 35, 55, 78, 79, 80, 87, 99, 130, 140, 148, 150, 164, 165, 181, 182, 183.
Woomera township/ Woomera Village, iv, v, 12, 13, 44, 56, 61, 74, 79, 84.
WRE Project Studies Group, 37.

X

X-51 Waverider (experimental vehicle), 88.
XMM-Newton (spacecraft), 68, 185.

Y

Yamba's Playtime, 96.
Yarragadee (WA), 66.
Young, Alan Hunter, 3, 4, 5.

Z

Zodiac (rocket), 1, 4.

www.ingramcontent.com/pod-product-compliance
Ingram Content Group UK Ltd.
Pitfield, Milton Keynes, MK11 3LW, UK
UKHW061344200426